Environmental Justice in Early Victorian Literature

This innovative new book combines environmental justice scholarship with material ecocriticism to explore the way in which early Victorian literature (1837–60) responded to the growing problem of environmental injustice.

As this book emphasises, environmental injustice – simply, the convergence of poverty and pollution – was not an isolated phenomenon, but a structural form of inequality; a product of industrial modernity's radical reformation of British society, it particularly affected the working classes. As each chapter reveals in detail, this form of environmental inequality (or 'classism') drew sharply critical reactions from figures as diverse as Thomas Carlyle, Friedrich Engels, Charles Dickens, and John Ruskin, and from within the Chartist movement, as working-class writers themselves reacted to the hazardous realities of a divided society. But as this book also reveals, these writers recognised that a truly just society respects the needs of the nonhuman and takes account of the material world in all its own aliveness; even if only tentatively, they reached for a more inclusive, emergent form of justice that might address the social and ecological impacts of industrial modernity, an idea which is no less relevant today.

This book represents an indispensable resource for scholars and students working in the fields of Victorian literature, environmental justice, and ecocriticism.

Adrian Tait is a UK-based independent scholar and ecocritic with a particular interest in Victorian literary responses to the impact of industrial modernity, and its relationship to questions of environmental and ecological injustice.

Routledge Studies in Environmental Justice

This series is theoretically and geographically broad in scope, seeking to explore the emerging debates, controversies and practical solutions within Environmental Justice, from around the globe. It offers cutting-edge perspectives at both a local and global scale, engaging with topics such as climate justice, water governance, air pollution, waste management, environmental crime, and the various intersections of the field with related disciplines.

The *Routledge Studies in Environmental Justice* series welcomes submissions that combine strong academic theory with practical applications, and as such is relevant to a global readership of students, researchers, policy-makers, practitioners and activists.

Environmental Justice in the Anthropocene
From (Un)Just Presents to Just Futures
Edited by Stacia Ryder, Kathryn Powlen, Melinda Laituri, Stephanie A. Malin, Joshua Sbicca and Dimitris Stevis

John Rawls and Environmental Justice
Implementing a Sustainable and Socially Just Future
John Töns

Intergenerational Challenges and Climate Justice
Setting the Scope of Our Obligations
Livia Ester Luzzatto

Solar Technology and Global Environmental Justice
The Vision and the Reality
Andreas Roos

Environmental Justice in Early Victorian Literature
Adrian Tait

For more information about this series, please visit: www.routledge.com/Routledge-Studies-in-Environmental-Justice/book-series/EJS

Environmental Justice in Early Victorian Literature

Adrian Tait

Routledge
Taylor & Francis Group
LONDON AND NEW YORK

earthscan
from Routledge

First published 2024
by Routledge
4 Park Square, Milton Park, Abingdon, Oxon OX14 4RN

and by Routledge
605 Third Avenue, New York, NY 10158

Routledge is an imprint of the Taylor & Francis Group, an informa business

British Library Cataloguing-in-Publication Data
A catalogue record for this book is available from the British Library

Library of Congress Cataloging-in-Publication Data
Names: Tate, Adrian, author.
Title: Environmental justice in early Victorian literature / Adrian Tate.
Description: London; New York: Routledge, 2024. | Series: Routledge studies in environmental justice | Includes bibliographical references and index.
Identifiers: LCCN 2023010507 (print) | LCCN 2023010508 (ebook) | ISBN 9780367420789 (hardback) | ISBN 9781032547664 (paperback) | ISBN 9780367821609 (ebook)
Subjects: LCSH: English fiction–19th century–History and criticism. | Environmental justice in literature. | Environmental protection in literature. | Human ecology in literature. | Ecocriticism.
Classification: LCC PR468.E58 T38 2024 (print) | LCC PR468.E58 (ebook) | DDC 820.9/353–dc23/eng/20230512
LC record available at https://lccn.loc.gov/2023010507
LC ebook record available at https://lccn.loc.gov/2023010508

ISBN: 978-0-367-42078-9 (hbk)
ISBN: 978-1-032-54766-4 (pbk)
ISBN: 978-0-367-82160-9 (ebk)

DOI: 10.4324/9780367821609

Typeset in Sabon
by Deanta Global Publishing Services, Chennai, India

To the memory of my parents.

Contents

Preface

The Victorians were living at a remarkable moment in human history, when industrial modernity began to generate (or intensify) the headline environmental problems that now capture the collective imagination, such as anthropogenic climate change. At the same time, however, industrial modernity was also creating more localised, more obvious harms, such as pollution-filled rivers and smoke-filled skies, and the burden of these harms fell primarily on the poor and disadvantaged. This was environmental injustice, as it is now termed, and it did not go unnoticed: in poetry and prose, the writers of the day responded to what they saw as a significant issue. Their work forms the subject of this book, which draws together environmental justice scholarship and ecocriticism in an 'ecojustice materialism', which it then uses to explore some of the various ways in which critics and commentators reacted against environmental injustice, inequality, and classism, and sought to conceptualise a solution. As this book illustrates, environmental injustice has a larger history, as part of a wider, structural shift, whereby a developing, modern society – a society within whose constraints we still live – generated patterns of inequality that encompassed race, class, gender, and ultimately, species; each of these categories has become the basis of forms of discrimination made manifest 'in the disproportionate burden of environmental harms facing these populations' (Pellow 2018, 5). In turn, and as this book also highlights, the Victorians began to explore a more expansive notion of justice – an 'emergent' justice – that also encompassed the nonhuman and more-than-human world, as themselves an integral and inherently valuable part of a fair and flourishing society.

*

Work cited

Pellow, David Naguib. 2018. *What is Critical Environmental Justice?* Cambridge: Polity.

Acknowledgements

I have now been writing and thinking ecocritically for some years, and many of the ideas that inform this book were first rehearsed at conferences and developed in essays and articles or in individual chapters in edited collections. I owe a great deal to all those scholars who have engaged with my work over the years, whether informally at conferences, or more formally through the process of reading, evaluating, and editing my work.

It perhaps goes without saying that I am indebted to those who went before me. In 2015, Jesse Oak Taylor wrote an article asking, 'Where is the Victorian Ecocriticism?' His point was a valid one – even then there was very little of it – but a select handful of ecocritics anticipated us all: it was, for example, Professor John Parham who first recognised the distinctive importance of Victorian thinkers and writers, in 'Was there a Victorian Ecology?' (2002). In turn, these pioneers inspired my own research: in particular, Richard Kerridge's discussion of an 'Ecological Hardy' (2000) encouraged me to think ecocritically about Thomas Hardy's verse, the subject of my doctoral thesis, which became the basis of my career as an independent scholar. I should add that my interest in the Victorians goes back still further; it would be remiss of me not to mention the historian Dr John Pemble, whose riveting lectures on the period highlighted to me how deeply British society is still shaped by the Victorians, for good or ill.

Although this book was written with an academic audience in mind, it forms part of what I take to be a larger ecocritical project, whose purpose is to encourage readers to think critically about the relationship between culture and nature. It is, in other words, literary analysis with a green political agenda, and as such, it constitutes its own, modest form of activism. In that spirit, I have sought to construct a narrative which acknowledges and engages with ecocriticism's complex theoretical apparatus (an apparatus that is as much philosophical as it is literary), but does so in a way that is nevertheless accessible to all those with an interest in our current environmental predicament, its roots, and its literary representation. To that end, I have tried wherever possible to identify sources that are available to a wider reading public, who do not have access to the research resources associated with a university. Fortunately, there are now several open-access sources of material associated with the Victorians. Moreover, many works by the

Victorians themselves (now long out of copyright) are freely available in the public domain. As I continually find myself reminded, there is no substitute for engaging first-hand with their own, often appalled, invariably interesting responses to the impact of industrial modernity. The world the Victorians saw emerging around them is, after all, very much our own, with all the problems and challenges that it entails.

Finally, I am grateful to the editorial staff at Routledge, on whose patience I have repeatedly prevailed. Writing this book has involved challenges which none of us could have anticipated when I began work on it in 2019!

I note that this book contains Parliamentary information licensed under the Open Parliament Licence v3.0, available at www.parliament.uk/site-information/copyright/open-parliament-licence/ (accessed 4 January 2023).

Introduction

The Victorian experience of environmental injustice

In his last finished novel, *Our Mutual Friend* (1864–5), Charles Dickens introduces us to the character of Reginald Wilfer, a man altogether too unassuming for so grand a Christian name. First glimpsed on his walk home to 'the Holloway region north of London' (Dickens 1864–5 [1997], 41), R. Wilfer is the 'shabby-genteel' (Dickens 1839 [1995], 304) representative of a new kind of figure in nineteenth-century society. Like Charles Pooter, a later (1892) literary inhabitant of Holloway, R. Wilfer is a clerk, a suburbanite, and by virtue of both, a member of the middle classes – but only just. '[H]aving a limited salary and an unlimited family' (Dickens 1864–5 [1997], 40), his position is a perilous one. His creditors have just repossessed the sign on his front door; the door may be next. Readers would have found this funny, as Dickens intended them to, but the humour is tinged with pathos. R. Wilfer can afford to live in the suburbs, but between his work and his home lies a wasteland that reminds him of how far it is possible to fall in Victorian Britain.

> Between Battle Bridge and that part of the Holloway district in which he dwelt, was a tract … where tiles and bricks were burnt, bones were boiled, carpets were beat, rubbish was shot, dogs were fought, and dust was heaped by contractors. Skirting the border of this desert, by the way he took, when the light of its kiln-fires made lurid smears on the fog, R. Wilfer sighed and shook his head.
>
> (Dickens 1864–5 [1997], 42)

This 'desert' was no invention. Dickens found the prototype for it at Belle-Isle, near King's Cross and St. Pancras (Dickens 1864–5 [1997], note 8, 808; see also Thornbury 1881, 276–8).[1] It was a landscape dominated by dust-heaps, mountains of cinders, ash, and household refuse often taller than the hovels they surrounded. Many such sites sprang up around London during the nineteenth century, all providing work for the waste-picker. Like the modern-day 'reclaimers' or 'diverters' of the developing world, from Bogota to Nakuru and Dhaka, these waste-pickers were the poorest of the poor (WIEGO n.d.).[2] As Dickens's journal, *Household Words*, reported in an article entitled 'Dust; or Ugliness Redeemed' (1850, 379–84), waste-picking

DOI: 10.4324/9780367821609-1

was the last resort of the unemployed and unemployable, of the old and the young but invariably of the infirm; of Peg Dotting, 'a poor old woman with a wooden leg'; of Gaffer Doubleyear, body 'nearly bent double'; and of 'little Jem Clinker, a poor deformed lad whose back had been broken when a child' ('Dust' 1850, 379). Amongst the dust-mounds, society's cast-offs gather to pick over its detritus, scraping a living by selling on 'soft-ware' (everything from 'a decayed cabbage-leaf' to dead cats, their pelts highly prized) and 'hard-ware' (broken pots, pans, earthenware, and even oyster shells) ('Dust' 1850, 380). They are 'the pariahs of the metropolis' (Thornbury 1881, 278). Trapped by their poverty, they have nowhere else to go: in a society that sees poverty as itself a kind of infection (Hilton 2006, 581), the only alternative is the workhouse, to which waste-picking the dust-mounds is preferable.

Ironically, the dust itself has a value that the waste-pickers do not. Cinders and ash, which make up the bulk of these mountains of dust, could be sold on to make bricks; in *Our Mutual Friend*, it is the brick- and tile-makers whose kiln-fires light up the lurid underworld that R. Wilfer traverses. As London itself expanded, '[e]normous quantities' of dust and ash were used in its building ('Dust' 1850, 382). Fortunes were made: one dust-heap, reported *Household Words*, was 'sold for forty thousand pounds, and … exported to Russia to rebuild Moscow' ('Dust' 1850, 384; see also Thornbury 1881, 278). In *Our Mutual Friend*, Old Harmon ('the growling old vagabond') has grown rich on dust (Dickens 1864–5 [1997], 24). When he dies, his employee Boffin (to whom his fortune defaults) suddenly becomes 'The Golden Dustman' (Dickens 1864–5 [1997], 137), setting up what Leslie Simon bluntly identifies as the 'controlling psychological economy' of the novel: '[m]oney lust' (2011, 221).

In a rapidly expanding and increasingly industrialised city, heavily dependent on coal for power and warmth, dust-mounds were neither a temporary nor an isolated phenomenon. When the investigative journalist James Greenwood visited Belle-Isle in 1873, he found that it was still home to 'several most extensive and flourishing dust-yards' (Greenwood 1873, 65–6), where 'scores of women and young girls find employment in sifting the refuse of dust-bins standing knee-high in what they sift' (65). But as Greenwood added, Belle-Isle was also home to 'bone boilers, fat-melters, "chemical works", firework makers, [and] Lucifer-match factories' (1873, 65). It was home, in other words, to 'almost every trade banished from the haunts of men' (Greenwood 1873, 65). All these poorly-paid, unpleasant, and sometimes dangerous trades – trades whose 'villainous smells' and 'dangerous atmosphere' no well-to-do neighbourhood would ever tolerate (Greenwood 1873, 65) – have simply moved to Belle-Isle. As Peter Darley points out, '[u]nsavoury businesses from tripe factories to horse slaughterers established themselves in the area, alongside breweries, tile kilns and chemical industries, adding to the gasworks (gas, tar, coke processing), coal depots, and open sewers' (2018–19, 10). This was accepted practice. Were you to complain to

Belle-Isle's manufacturers, the architect George Godwin observed in a treatise tellingly entitled *Town Swamps and Social Bridges* (1859):

> ... you will probably be laughed at, and told, that 'This is a *nuisance neighbourhood*. What business have you to interfere? Leave us alone – we neither hurt ourselves nor anybody else'. Let us, however, say in reply, that what seems sport to him, is death to others.
>
> (Italics in the original; Godwin 1972, 11)

As Godwin insisted, these businesses do have an impact, but only on those whom society dismisses as expendable. These 'nuisance neighbourhoods' constitute what are today called 'ecological sacrifice zones' (Faber 2018, 62). Here, entire communities 'are marked for erasure and early death' (Faber 2018, 62), and subject to 'ideological and institutional othering' that positions them as unimportant, like 'the more-than-human world' itself (Pellow 2018, 17). For the self-interested manufacturers of Belle-Isle, the economic logic of the situation is impeccable: in an area such as this, they can draw on a workforce too poor to care about the work they do, and its disastrous effect on health. As Greenwood reported, 'swarms of little children and grown men and women abide winter and summer in this awful place ... It is utterly impossible that the poor wretches doomed to [live here] should not be afflicted' (1873, 67).

This, then, is the sequestered corner of the city that R. Wilfer glimpses as he trudges homewards through the hinterland between workplace and suburb, himself too poor to afford the omnibus that might spare him these sights:

> 'Ah me!' said he, 'what might have been is not what is!'
> With which commentary on human life, indicating an experience of it not exclusively his own, he made the best of his way to the end of his journey.
>
> (Dickens 1864–5 [1997], 42)

It is that 'commentary on human life' with which I am concerned in *Environmental Justice in Early Victorian Literature*. What might have been is not what is, nor what ought to be: in a variety of literary forms ranging from fiction-with-a-purpose to the emotive exposés of investigative journalists and family entertainments that sought to enlighten as well as entertain, Victorian writers responded to (and reacted against) the iniquities they saw around them. As they recognised, Belle-Isle was not an isolated example of what we would now call an environmental injustice, whereby a community is exposed to a localised and specific hazard or environmental harm: infamous as it became, Belle-Isle was simply a part of a wider pattern of what David Pellow has defined as environmental inequality, 'a sociohistorical process' rather than 'a discrete event' (2000, 588) which in turn demands a focus on 'the much deeper workings of power in society' (597). Everywhere in Victorian Britain, an industrialising nation found it easier to dump its hazards on the dispossessed and disadvantaged, give them the worst kinds of work, and abandon them to the neighbourhoods its better-off abandoned and avoided. This was not simply a question of the burdens that society heaped on the poorest of the poor, the otherwise unemployable underclass to which *Household Words* referred. Using wages as an index, the historian K. Theodore Hoppen argues out that 'the line of poverty clearly ran *through* the working class' (italics in the original; 1998, 63), but even when wages were above subsistence level and employment *was* continuous, the working class could hardly be said to have flourished. In the world made by industrial modernity, impoverishment took on forms more subtle than hunger. In cities both old and new, the working classes lived in close proximity to the polluting factories in which they found (often hazardous) employment. They were the victims, in other words, of a form of inequality that Karen Bell describes as environmental classism (2020, 2), a term that links questions of environmental health and environmental justice to social stratification within industrialised societies.

As this brief discussion suggests, what we think of as 'environmental injustice' is not a new (nor an isolated) development. But just as environmental injustice has a past, so too does the struggle to resist it, as the literature of the period bears witness. Indeed, writers saw themselves as part of that struggle. For writers such as Thomas Carlyle, Friedrich Engels, John Ruskin, and Charles Dickens, and for the now-forgotten, often anonymous writers who contributed to the Chartist movement's literature of resistance, the aim was not just to describe or evaluate this emerging problem, but to address it. Nor were these writers exclusively concerned with alleviating isolated social inequalities; they also sought to reconceptualise the nature of society in ways that might also promote a flourishing, nonhuman world. As they sensed, even before the emergence of ecology (as science) and environmentalism (as political movement), human lives are intimately caught up with and co-dependent on a nonhuman and more-than-human world, and one cannot flourish without the other.

How might we explore the literary response to this emerging issue, and do justice to its sense of the interrelationship between social and ecological concerns? As I discuss in more detail, below, this book combines two fields of study: environmental justice scholarship, or the field of 'EJ Studies' (Pellow 2018, 9), and ecocriticism, or the study of cultural constructions of nature and the environment; in particular, I draw on a form of ecocriticism rooted in new materialist theories, hence a 'material ecocriticism'. Environmental justice activism was (at least initially) preoccupied with the plight of marginalised human communities, and concerned with the environment only to the extent that it represented a source of harms or goods. By contrast, ecocriticism was at first concerned to reassert the importance of the nonhuman, natural world; arguably, it neglected the social, human dimension of environmental crisis. Over time, however, the two approaches have converged, as each has developed more expansive theoretical frameworks. It is now possible to align these approaches to develop an 'ecojustice materialism', whose parameters reflect the environmental justice movement's concern with the root causes of environmental injustice (notably the impact of industrial modernity), a new materialist concern with the co-constitutive relationship between human, nonhuman, and more-than-human entities, and an ecocritical focus on the material *and* discursive dimension of those interactions.

From this hybridised perspective, the importance of exploring Victorian attitudes and experiences becomes clear. On the one hand, the period marked the widespread emergence of environmental inequality and classism, as industrial capitalism combined with a modernising, progressive narrative to transform society's structural basis; it is in this sense that I use the term 'industrial modernity', to denote the convergence between what Ulrich Beck describes as 'the industrial-capitalist colossus' and '[t]he ensemble of identifications … industry = progress = science = enlightenment = modernity' (1995 [2002], 5). On the other, and as part of the complex dialectic that characterises the period, it was defined by a questioning, critical attitude towards the social *and* ecological impacts of this transformation, as new ideas and discoveries transformed attitudes towards the nonhuman world. The result was a substantial body of literature, spanning poetry and prose, fiction and nonfiction, engaged in often openly critical discussions about the nature of modern society, and the direction it should take. As I explain in this Introduction, my aim is to analyse a selection of texts that reflect this critical diversity of opinion about what constitutes environmental justice, focusing on the period from 1837 to 1860, when these concerns were felt most acutely, and the literary response was at its most engaged and urgent.[3] These literary responses have a continuing relevance today, as societies grapple with the interlinked impacts (at once social and environmental) of industrial modernity's rise to global dominance.

*

The environmental justice movement emerged in the United States in the 1980s as a grassroots response to the dumping of toxic wastes on impoverished communities (Bell 2014, 15; Bullard, Johnson, and Wright 1997, 68; Buell 2005, 117). Since then, similar struggles around the world have combined to form a global movement, united behind the belief that in a just society 'all people … share equally in the benefits bestowed by a healthy environment' (Adamson, Evans, and Stein 2002, 4). For many, however, day-to-day reality is shaped by environmental discrimination, or the '"unequal treatment" of a group based on race or class' (Bullard, Johnson, and Wright 1997, 64). As Ryan Holifield et al. point out, '[e]nvironmental problems, from water and pollution to biodiversity loss and global warming, have the capacity to affect us all', but 'they do not affect us all equally, or in the same ways' (2018, 1). As they add, '[t]his unequal and differentiated positioning, which typically places the heaviest environmental burdens upon the marginalised, disadvantaged, and less powerful populations, forms the central premise of the problem of environmental injustice and the hope for environmental justice as its solution' (Holifield, Chakraborty, and Walker 2018, 1).

As the environmental justice movement has expanded, so too has its scope, raising the possibility that environmental justice should also encompass the nonhuman world. 'The shift suggested here', notes David Schlosberg, 'is one from environmental conditions as an example or manifestation of social injustice to one where justice is applied to the treatment of the environment itself' (2013, 44). The importance of this shift is underlined by Timothy Clark, who argues that focusing on 'the destructive effects of human systems of hierarchy and inequality' does not automatically address environmental issues on the scale of (for example) of climate change (2011, 74). Nor does it address the fundamental problem, of a natural world that is still subject to the 'dualistic separation of humans from nature promoted by Western philosophy and culture' (Garrard 2012, 24), and valued only because of its utility, as a source of harms or goods to be divided up by society.

How might this shift or extension be achieved? With the growth of 'EJ Studies', scholars have developed more detailed understandings of what is meant by 'environmental justice'. 'In addition to *distributive* justice', explain Holifield et al., 'a growing body of literature now attends to *procedural* and participatory justice, justice as *recognition*, and justice as *capabilities*, as well as the interrelations among these dimensions' (italics in the original; 2018, 4). In turn, Schlosberg argues, it is 'possible to use this expanded set of justice discourses ... when discussing relationships of justice between the human and nonhuman realms'; 'indeed, the same conceptions can be applied to both environmental and ecological matters' (2007, viii). Take, for example, the distributive and procedural approaches to environmental injustice, according to which injustice is a function of 'the distribution of environmental goods and harms' and of a lack of fairness in 'the process of environmental policy- and decision-making' (Bell and Carrick 2018, 101). Both distributive and procedural paradigms take as their basis 'the widely accepted notion that people deserve equal treatment' (Whyte 2018, 113). It follows that, for nonhuman nature to receive the same equality of treatment, it must itself merit the same moral consideration. One approach to this problem – the problem of establishing moral parity – is to recognise and acknowledge 'the similarities between human beings and the rest of nature' (Schlosberg 2007, 158). Another is suggested by the recognition paradigm. By contrast with the distributive and procedural paradigms, the recognition paradigm is concerned with the way in which environmental injustice emerges because of a failure 'to respect or acknowledge difference' ('as opposed to sameness or strict equality') (Whyte 2018, 113). In other words, the recognition paradigm shifts the focus away from 'violations of sameness, such as equal treatment before the law', to societal or systematic failures 'to acknowledge or recognise *difference*' (italics in the original; Whyte 2018, 117). It is not on the basis of sameness that moral (and hence legal or juridical) recognition is warranted, argues Kyle Whyte, but on the basis that 'plants, animals, insects and ecosystems are also implicated in EJ situations' (2018, 119).

Whyte's argument hinges on the idea that human flourishing and nonhuman ecological processes are so intimately entangled that there can be no strict separation of one from the other. Given that humans and nonhuman communities are dependent on and affected by each other, justice must be applied to both (Schlosberg 2013, 44), underlining the need for what Schlosberg calls 'participatory parity' (2007, 158). This requires a deliberative effort to expand the basis of the polity. As scholars argue, uneven exposure to risk 'is symptomatic of a lack of democracy' (Faber 2018, 61), since democratic participation provides the means of responding to those risks. It follows that ecological justice depends on the 'ecological extension of the familiar idea of a democracy of the affected' (Eckersley 2004, 112); it requires, in other words, the deliberative 'recognition and inclusion of nature in an egalitarian democratic politics dedicated to transcending and helping to breakdown the boundary between human beings and nonhuman nature' (Schlosberg 2007, 192).

As these arguments highlight, a just society is one that takes into account the flourishing of human and nonhuman species and the functioning of more-than-human systems. Creating that society is, however, impossible without first pinpointing the root causes of environmental injustice (Bullard, Johnson, and Torres 2005, 294). Karen Bell identifies several causal explanations, including discrimination, market dynamics, and a lack of citizen power (Bell 2014, 33), but what all these explanations have in common is a recognition that industrialisation is a significant contributory factor: environmental injustice is often linked to 'ecologically hazardous industries', and the environmental justice movement itself emerged in response to the siting of 'hazardous industrial facilities and waste sites' in poor communities (Faber 2018, 63). As Bell adds, industrialisation 'is associated with extensive environmental degradation, as well as threats to human health from occupational hazards and waste production' (2014, 39). Furthermore, and whilst the case is often made for the benefits that industrialisation may bring, 'research casts doubt on the idea that there is a simple relationship between industrialisation, development and progress' (Bell 2014, 40). The Industrial Revolution brought 'immense wealth for some', Bell notes, but significant costs to others, exacerbating 'the plight of poor and marginalised people' (2014, 40) whilst also generating new and unexpected environmental impacts, ranging from the local and immediate to the regional, global and longer term, from acid rain to global warming and ocean acidification.

Industrialisation is, therefore, a key component in what we now think of as environmental injustice, as the convergence of 'poverty and pollution' (Bullard, Johnson, and Torres 2005, 294). It also lies behind what is now widely described as the 'Anthropocene' (Crutzen and Stoermer 2000), a geological epoch permanently marked by human activity, and scarred by often 'incremental and accretive' forms of what Rob Nixon describes as 'slow violence' (2011, 2). Yet industrialisation could not have occurred without the conditions created for it by capitalism, which Bell also considers 'a root determinant of environmental injustice' (2014, 33). We can be more specific, and point to unregulated, free-market forms of capitalism, which characterised Western economies such as Britain's at the time of its Industrial Revolution, and dominate those economies again today: as John F. Weeks points out, today's neoliberal arguments are simply a version of those that prevailed 150 years ago (2014, 192). Thus, free-market economies are allowed to predominate, and since 'a market economy requires profit and growth ... these priorities come to dominate decision-making at the expense of ecological and social concerns' (Bell 2014, 215). In the absence of regulation or control, capitalism fosters competition, and as Weeks contends, 'unregulated competition disintegrates society into alienated and mutually suspicious individuals, and *de facto* divides these individuals along class lines' (2014, 190–1).

In turn, argues Daniel Faber, capitalism registers 'social losses' such as air pollution as 'negative externalities' (2018, 61) that have no fundamental bearing on capitalism's continued functioning. Capitalism simply seeks out ways of 'displacing negative externalities that are simultaneously the most *economically efficient* and *politically expedient*' (italics in the original; Faber 2018, 61). In seeking 'the path of least political resistance' (Faber 2018, 62), capital therefore targets the most disempowered and disadvantaged communities. Thus, and whilst Faber acknowledges that 'there are multiple political-economic forces at work that give [each] injustice a particular context and form', he contends (like Bell) that 'environmental injustices are rooted in processes of capital accumulation' (Faber 2018, 62).

Coupled with industrialisation, the self-serving work of capital goes some way to explaining the existence of environmental injustice in the modern world, and more broadly, humankind's escalating environmental impact; on this basis, argues Jason W. Moore, the term Anthropocene (which implies 'humanity as an undifferentiated whole') should be replaced by 'Capitalocene, understood as a system of power, profit and re/production in the web of life', and by an emphasis on 'historical capitalism as a world-ecology of power, capital and nature, dependent on finding and coproducing Cheap Natures' (2017, 2).[4] Yet the mainstream environmental justice movement has fought shy of recognising the explanatory power of capitalism as a dominant (perhaps even predominant) driver of environmental inequality. Although many of its defining struggles were fought in the 1980s, Nancy Fraser has argued, the environmental justice movement came to maturity in a 'postsocialist' aftermath (1997, 1) whose sceptical mood called into question the possibility of 'an alternative to the present order' (2). In the absence of an overarching, progressivist vision of an alternate, post-capitalist future, scholars suggest, oppositional movements retreated into forms of identity politics that eschewed broader questions about social equality and political economy (Parenti 1997; Dean 2012), a point that recent research drives home (Pulido 2018; Bell 2014).[5] In the US, Laura Pulido argues, the environmental justice movement 'has become far less oppositional' (2018, 17), and its approach 'largely devoid of significant material demands' (21). As Bell adds, 'mainstream use of the term' 'Environmental Justice' has increasingly tended to avoid its more radical implications, restricting its definition in a way that avoids any challenge to economic growth (2014, 16). In this formulation of what constitutes environmental justice, the market is regarded as 'the best solution' (Pulido 2018, 21).

Whilst '[t]heories of discriminatory intent seem to ignore structural processes' (Bell 2014, 63), however, capitalism alone does not explain discrimination. Capitalism may exploit and deepen existing inequalities, both economic and political (Weeks 2014, 187), but its role in creating those inequalities is more obscure. Certainly, capitalism requires the existence of a class structure, and it assumes and may even foster the creation of a propertyless and powerless working class with no recourse but to sell its own labour, but arguments for the role of capitalism do not offer an adequate explanation for discrimination on the grounds of (say) race or gender. It also follows that, whilst the market may worsen discrimination, intervention in the market does not 'eliminate or even substantially reduce the economic effects of discrimination' (Weeks 2014, 201).

Where, then, should we turn for an understanding of the discriminatory dimension of environmental injustice? Bell focuses on the role played by hegemonic discourses that present themselves as natural and inevitable ways of thinking about the world: 'alongside the structural constraints of capitalism, [they] create the confining conditions within which agencies operate' (Bell 2014, 10). Bell identifies a cluster of related discourses that 'shape and support our culture and values and, in turn, our actions' (2014, 10), and more specifically, sustain environmental injustice. Of these damaging and hegemonic environmental discourses, Bell argues,

> [t]he most relevant, taken-for-granted … appear to be the beliefs that ever-expanding growth is 'good' or 'necessary'; that the environment is separate from humans and needs to be controlled (or 'mastered'); that the environment is of minimal importance for human health; and that complex, modern or 'high' technology is preferable to more basic or traditional technology.
>
> (2014, 9)

As she also notes, these hegemonic perspectives assume that '[t]here is a linear rationality to the universe' (Bell 2014, 43).

These interlocking arguments can, in turn, be traced to what Val Plumwood has described as 'the great empiricist transition of the Enlightenment', the creation of a human-centred and 'hegemonic rationality' (2002, 15), and the model of mastery it has promoted. As she argues in *Feminism and the Mastery of Nature* (1993 [1997]), this hegemonic rationality created a dominant and still prevalent belief in the dualistic separation – or 'hyperseparation' (1993 [1997], 49) – of the reasoning human mind (and the culture that was its product) and the unreasoning world of nature, a world further distinguished by its passivity, its pliability, and its predictability. There is, Plumwood argues, a double movement at work in this process of hyperseparation, which permits a 'logic of colonisation' (1993 [1997], 41) from which environmental crisis has itself sprung. On the one hand, (nonhuman) nature is separated out as distinct from (human) culture; on the other, nature is defined as inferior to it, and as such, freely available for human exploitation. This rationalising, instrumentalising, reductionistic spirit came to underpin (scientific) 'modernity and manipulative technoscience' (Plumwood 2002, 47), and at last make possible the 'historical projects of subduing and colonising nature' (15). The 'rationalist hyperseparation of human identity from nature' (2002, 9), Plumwood continues, is a pivotal factor in creating environmental crisis, a crisis (she argues) 'of a cultural "mind" that cannot acknowledge and adapt itself properly to its material "body", the embodied and ecological support base it draws on in the long-denied counter-sphere of "nature"' (15).

There is, however, another dimension to this process, and its 'construction of a devalued and sharply demarcated sphere of otherness' (Plumwood 1993 [1997], 41). The opposition between culture and nature is, in fact, simply one of a 'set of interrelated and mutually reinforcing dualisms which permeate [W]estern culture', 'a fault line which runs through the entire conceptual system' (Plumwood 1993 [1997], 42). Just as hyperseparation creates a double movement whereby (nonhuman) nature is separated out as distinct from and inferior to (human) culture, so through a succession of related dualisms, certain groups within human society are themselves identified as inferior to the rational, white, male norm because supposedly closer to that nature. These dualisms variously operate to 'naturalise gender, class, race and nature oppressions' (Plumwood 1993 [1997], 43).[6]

Plumwood's compelling argument for the constitutive features of this 'modern, post-enlightenment consciousness' (1993 [1997], 43) – and its discourse of modernity – provides an understanding of the ideological bases of multiple forms of discrimination 'centred on class, gender, race, [and] ethnicity', which capitalism itself exploits (Faber 2018, 62). It is on the basis of exclusionary processes such as these, argues Faber, that capitalism operates to impose 'ecological sacrifice zones' on the 'subaltern' (2018, 62). The subaltern are, he explains, the 'effectively devalued' component of a society, who consequently experience 'multiple forms of political domination, economic exploitation, and cultural oppression' (Faber 2018, 62).

Whilst 'no single causal factor seems to completely explain environmental injustice' (Bell 2014, 63), therefore, there is a strong case for the threefold role of capitalism, the process of industrialisation, and modernity's dualistic model of mastery. Environmental injustice emerges, in other words, from the interplay between industrial capitalism and modernity's hegemonic discourse, a discourse built on the discriminatory, dualistic logic described by Plumwood. This convergence of capitalism, modernity, and industrialisation is, in turn, characterised by its global reach, restless expansionism, often accidental harms, and disregard for all those that it marginalises or externalises.

Along with the interpretative paradigms discussed earlier (distributive, procedural/participatory, justice as recognition or capabilities), the idea of this causal convergence provides a useful tool for analysing the emergence of environmental injustice, better understanding its continued existence, and framing responses to it. For example, it highlights the fact that environmental inequality has historical roots in structural forces of the most fundamental kind. The complex, interlocking nature of these problems underlines the difficulty of finding solutions that might lead to a just society, a society in which justice extends to humans and nonhumans, and which recognises the needs of those future generations that may be affected by the slow violence to which Nixon refers. If the problem is capitalism, it follows that no solution is possible that does not involve its thorough-going reform or comprehensive regulation, or as its sternest critics suggest, a revolution to supplant it; if the problem is and continues to be industrialisation, then a just society must be post-industrial; if the problem lies in an environmentally damaging but also discriminatory discourse that has now been internalised and naturalised within Western culture, then this too must be exposed, interrogated, and countered. These are immense challenges, and they underline the need to consider radical as well as reformist solutions – or perhaps radical rather than reformist solutions – if the environmental justice movement is to move beyond grassroots struggles with localised instances of environmental injustice and make good on its initial promise to engage with deeper, structural environmental inequalities.

At the rather more modest level of an academic study concerned with the way in which Victorian literature might also help us to a better understanding of contemporary environmental inequality, two points follow. First, the search for environmental justice necessarily demands a historical understanding of the social, economic, and political forces that drive injustice. As environmental justice scholars acknowledge, 'historical processes provide insights into the processes that create environmental injustices in the present as well as the past', and help 'elucidate the multiple factors and complexities that generate uneven environmental consequences for different groups of people' (Boone and Buckley 2018, 228). It follows that environmental justice studies should take into account the epochal shifts experienced by the Victorians, at a moment in human history when society was for the first time being reshaped by the entirely novel convergence of capitalism, industrialisation, and modernity.

Second, it is equally important to engage with the discursive dimension of environmental injustice, and its roots in culture. This matters not only because discourse/culture plays a formative (and performative) role in creating and perpetuating environmental injustice, but because, as environmental justice scholars have also emphasised, art and literature have a positive role to play in the continuing struggle for environmental justice. Making art, argue Joni Adamson et al., is itself an 'intrinsically political' act (2002, 7), and as Julie Sze adds, literature can offer a 'more flexible representation of environmental justice, one with a global view and historical roots', freed 'from providing a strictly documentary account of the contemporary world' (2002, 163).

This leads us to the role that ecocriticism – or environmental criticism – might itself play in expanding environmental justice scholarship. Ecocriticism focuses on the relationship between the nonhuman, natural world, and its representation in culture throughout human history (Garrard 2012, 5). As such, and as itself 'an avowedly political mode of analysis' (Garrard 2012, 3), it is well suited to exploring the discursive, cultural dimensions of environmental injustice, and deepening our historical understanding of its emergence. Motivated by a 'troubling awareness that we have reached the age of environmental limits' (Glotfelty 1996, xx), ecocriticism first emerged in the US in the 1990s, where it was defined very simply as 'the study of the relationship between literature and the physical environment' (Glotfelty 1996, xviii). From the outset, ecocriticism considered itself part of a broader environmentalist praxis: as one of its pioneers, Cheryll Glotfelty, insisted, 'literature does not float above the material world in some aesthetic ether, but, rather, plays a part in an immensely complex global system, in which energy, matter, *and ideas* interact' (italics in the original; Glotfelty 1996, xix). Initially, this 'first-wave' of US ecocriticism drew its inspiration from the kind of deep ecological thinking (Parham 2010, 31–4) that Ramachandra Guha and Joan Martinez-Alier would later characterise as the '"full-stomach" environmentalism of the [Global] North', an 'environmentalism [conceived of] in largely mental terms as a question of values affirmed or denied, "post-materialist" or "anti-materialist"' (Guha and Martinez-Alier 1997, xxi). According to Pippa Marland, deep ecological thinking 'asserts that if we first address our hierarchical attitudes towards the natural world and identify ourselves within a broader circle of living things, then our societal problems may also find resolution' (2013, 850). As such, it emphasises the role of the individual – and of 'an individual adjustment of values' – in 'a radical reconceptualisation of humanity's place on the planet', and adopts the ecocentric valuation of nature as important in itself and not just as 'a resource for humanity' (Marland 2013, 850).

This first-wave, deep ecological emphasis was, however, problematic in its own right. Although deep ecology was and remains a significant influence on ecocritical thinking, particularly in the US, it has been criticised for its faith in 'individual enlightenment' as an agent of social change, and its reluctance to engage with the 'socio-political factors' (such as 'global capitalism' and 'the ideology of consumerism') that brought about environmental crisis in the first place (Parham 2010, 45). In turn, ecocriticism's first-wave focus on wilderness and wilderness preservation was criticised as the expression of a dematerialised and 'essentially contemplative privileged attitude to the natural world' (Clark 2011, 87). Indeed, the environmental justice movement itself reacted against what it perceived to be ecocriticism's failure to take the social dimensions of environmental crisis seriously. As T. V. Reed wrote in 2002, '[t]he lack of a strong environmental justice component within the field of ecocriticism should be felt as a deep crisis' (157).

In calling for a new kind of 'Environmental Justice Ecocriticism' (Reed 2002, 149), Reed nevertheless conceded that work was already 'well underway' (160). As it took a 'more sociocentric direction', notes Lawrence Buell (2005, 112), ecocriticism began to engage with the theories and practices of the environmental justice movement, recognising shared areas of concern in its own emerging interest with 'toxification narratives' and 'thematic configurations' such as 'eco-apocalypticism' and 'environmental racism' (2005, 130). Buell's own work reflects that shift. In *Writing for an Endangered World* (2001), for example, he discussed and anatomised the emergence of forms of 'toxic discourse' in European and American literature, and their manifestation in nineteenth-century accounts of 'Gothicised environmental squalor', 'perhaps the best known being Friedrich Engels's description of Manchester and other British factory towns in *The Condition of the Working Class in England* (1845) and Charles Dickens's novel *Hard Times* (1854), also set in the industrial Midlands' (Buell 2001, 43). Returning to the same subject matter in 2005, Buell discussed the existence of the 'environmental illness narrative' (2005, 119) – Engels's *The Condition of the Working Class* was, again, 'the classic documentary work' (120) – and a literature of 'ecological refugeeism' that could be traced to 'the British enclosure movement' of the late-eighteenth century (121).

As these examples suggest, Buell sought to open up an approach that combined environmental justice concerns with an ecocritical emphasis on the representation of environmental injustice in literature, a hybrid approach that (in order to distinguish it from first-wave ecocriticism) he labelled an 'environmental justice revisionism', or for short, 'ecojustice revisionism' (Buell 2005, 112). Given that deep ecological thinking remained an influential part of the ecocritical project (2005, 126), however, Buell acknowledged that ecojustice revisionism would 'be criticised … as restricting ecocriticism's agenda overmuch to issues of social justice and equity' (2005, 127). The difficulty, Buell foresaw, was the extent to which environmental justice scholarship retained its initial, US view of the 'environment' as important only to the extent that it represented a source of goods or harms to be divided up more or less equitably between different sections of the community, a view that was essentially at odds with 'one of ecocriticism's founding gestures … to reject the adequacy of reading "nature" as no more than a function of cultural politics' (Clark 2011, 93).

Perhaps as a result, there has (until recently) been little ecocritical interest in Buell's putative new direction,[7] or its application to Victorian culture, despite Buell's own emphasis on the importance of nineteenth-century literature. Victorian ecocriticism may now be 'an established and accepted practice', as Laurence Mazzeno and Ronald Morrison contend (2017 [2019], 5), but relatively little of it has thus far pursued a sociocentric focus on issues of environmental equity.[8] In this respect, essays such as John Parham's 'Bleak intra-actions: Dickens, turbulence, material ecology' (2017 [2019]), edited collections such as Dewey W. Hall's *Victorian Ecocriticism: The Politics of Place and Early Environmental Justice* (2017), and monographs such as Lance Newman's *The Literary Heritage of the Environmental Justice Movement* (2019) are the exception rather than the norm.[9]

Whilst ecocriticism has been slow to engage with 'EJ Studies', however, it has continued to explore other new directions – indeed, the 'waves' of ecocriticism's development may now have multiplied to the point where the descriptor is all but meaningless – and these new directions also have a sociocentric dimension. In recent years, for example, ecocriticism has taken a growing interest in ecofeminist and postcolonial arguments, which are also concerned with the 'inequitable distribution of environmental benefits and risk among the global population' (Marland 2013, 852). No less significantly, ecocritics have turned to new materialist theories, which include, amongst others, Stacey Alaimo's concept of 'trans-corporeality, in which the human is always intermeshed with the more-than-human world' (Alaimo 2010, 2), Jane Bennett's 'vital materialist vocabulary' of '[t]hing power', or the power of supposedly passive, inert matter 'to manifest traces of independence or aliveness' (Bennett 2010, xvi), and perhaps most notably, Karen Barad's ground-breaking theory of 'agential realism', which highlights the nature of 'matter as a dynamic and shifting entanglement of relations' (Barad 2007, 35). In Barad's subject–object blurring reading of the continuous and constitutive entanglement of entities – a reading that blurs the subject–object divide – the human body is reinstated as itself a part of the natural world, and as itself natural. As Alaimo contends, it is vital to acknowledge the 'material interconnections of human corporeality with the more-than-human world' (2010, 2), and in so doing, collapse the dualistic distinction implied by 'nature' and 'culture' but also 'human' and 'environment' (2).[10] Furthermore, Barad's theory of agential reality emphasises 'the causal relationship between discursive practices and material phenomena' (Barad 2007, 34) – of 'naturalcultural practices' (32) and 'material-discursive factors (including gender, race, sexuality, religion, and nationality, as well as class)' (35) – and in so doing offers a more expansive way of taking into account 'the agential possibilities and responsibilities for reconfiguring the material-social relations of the world' (35).

New materialist theories have already influenced feminist thinking (thereby generating 'material feminisms'; Alaimo and Hekman 2008), ecocriticism (hence 'material ecocriticism'; Iovino and Oppermann 2014), and feminist ecocriticism (Oppermann 2013), but they also suggest both common ground with – and productive ways in which to reinterpret – environmental justice

(Marland 2013, 855) and environmental health concerns, which themselves 'mark significant interchanges between human bodies and [toxic] environments' (Alaimo 2010, 3). By extension, this overlap with the concerns of environmental justice scholarship also suggests a link between new materialism and more conventional, materialist conceptions of history from which it is theoretically distinguished; as we have seen, there is a strong case for arguing that environmental injustice is rooted in and an extension of the same forces of class exploitation to which Marxist historians have repeatedly drawn attention, in a critique that itself dates from the Victorian period. Indeed, grounding new materialist theory in the lived experience of environmental injustice may be a constructive response to Terry Eagleton's contention that, whilst new materialism 'emerged in part to replace a currently unfashionable historical materialism ... whole currents of it would seem to have no particular concern, as historical materialism does, with the destiny of men and women in an exploitative world' (2016, 17).

In addition, the new materialist acknowledgement that 'material interconnections' matter is a step towards 'more capacious epistemologies [and more expansive] ethical and political positions' (Alaimo 2010, 2), positions that overlap with more recent environmental justice thinking about an expanded notion of justice that takes into account nonhuman as well as human flourishing; with its acknowledgement of the need to acknowledge 'the more-than-human world (from nonhuman animals and ecosystems to the built environment) as subjects of oppression and as agents of social change' (Pellow 2018, 19), David Pellow's concept of a 'Critical Environmental Justice' is a case in point (2018). As Alaimo suggests, a new materialist 'sense of being embedded in, exposed to, and even composed of the very stuff of a rapidly transforming material world' suggests the need for new 'forms of ethics and politics' (Alaimo 2016, 1) that reflect this sense of hybridity and inseparability. As the dualistic separation between 'nature' and 'culture' collapses, it becomes increasingly difficult to justify an ethic that draws the line at humans and excludes nonhumans, or that thinks of environmental justice only in terms of impacts on human communities. To the contrary, new materialist theories such as Barad's suggest the need to conceptualise justice as an integral part of 'a dynamic and shifting entanglement of relations' (Barad 2007, 35) that encompasses matter and discourse, human and nonhuman, but also past, present and future (between which, Barad argues, '[t]here is no inherently determinate relationship' [2012, 66]). What this suggests, contends Barad, is a form of justice akin to 'Derridean notions of justice-to-come' (Barad 2012, 67), a justice '*à venir*' (Derrida 1992, 27) that (as Jacques Derrida put it) cannot, 'must not wait' (Derrida 1992, 26), but remains 'irreducibly to come' (27). It is, in other words, an evolving, emerging, and processual form of justice whose outcome we do not presume to know 'in advance and which is forever fixed' (Barad 2012, 67), but which is nevertheless animated by an urgent and unavoidable moral imperative to act.

The practical challenge of defining such an expansive but also evolving sense of justice is difficult to overstate; for the sake of simplicity, it might be called an 'emergent justice', to reflect the need for a processual, pragmatic approach to questions of justice, but also to distinguish it from both an 'environmental justice, which addresses environmental risks within human communities, and ecological justice, focused on the relationship between those human communities and the rest of the natural world' (Schlosberg 2007, 3). Barad's theory of agential realism collapses this distinction, just as it collapses the dualistic opposition of nature and culture on which it rests; communities are never finally separable from their environs, but are always already caught up in wider material flows that undermine the conventional distinction between environmental justice and ecological justice. 'Emergent justice' is, therefore, convenient shorthand for an alternative and expansive understanding of what constitutes the legitimate field of concern for justice advocacy.[11]

If, now, we combine environmental justice concerns with Buell's ecojustice revisionism, and more recent, new materialist inflexions of ecocriticism, we arrive at a fusion that might be called (at the risk of creating yet another neologism) 'ecojustice materialism'. Simply, ecojustice materialism is concerned with the representation of environmental injustice, the complex entanglements that constitute it within the field of naturalcultural practices, and with the imaginative conceptualisation of an emergent justice that might enable societies to look forward to truly just alternates. An ecojustice materialism also looks back, seeking out and interrogating the historical origins of environmental injustice as a material-discursive configuration driven by capitalism, shaped by industrialisation, and underpinned by hegemonic, divisive, and discriminatory discourses of modernity.

What this suggests, in other words, is a theoretical apparatus that enables scholars to analyse the material-discursive formation of environmental injustice, but also its structural corollary, environmental inequality, and its class-bound expression. What an ecojustice materialism may also prompt is a concern with the epochal changes that convulsed Victorian Britain, and caused so much doubt and dismay – even as it inspired such confidence and pride – amongst contemporaries. As the historian L. C. B. Seaman remarked, what happened here 'had never happened before; and once it had happened … there was hardly a people on earth who, a century later, had not caused it to happen to themselves or who were not strenuously trying to make it happen' (1973 [1993], 26). Victorian Britain was 'the first human society to pass into industrial modernity', notes Jesse Oak Taylor (2016, 4), and perhaps the first to live within the Anthropocene (Taylor 2015, 878). Underpinned by the hegemonic discourses that Bell describes, industrial modernity became a self-sustaining force during Victoria's reign, creating a model for future human development whose environmental impacts have only been fully recognised in recent decades, and with which humankind is only now wrestling; from the perspective of 'this present age of ecological emergency', argues Timothy

Morton, 'we are still inside the Victorian period, in psychic, philosophical, and social space' (2014, 489). But the Victorian period was also one in which this convergence of material forces and discursive factors began to create recognisably modern, global forms of environmental inequality. For much of the Victorian period, capitalism was allowed free rein, with minimal state regulation, and few limits on its exploitation of less developed countries. As such, the *laissez-faire* capitalism of Victorian political economy closely resembles the free-market neoliberalism that now dominates mainstream economic thinking in a globalised world (Harvey 2005 [2007], 3; Gould 2013, 5–8). Industrial innovation created new forms of environmental hazard; as noted earlier, these were simply dumped on poor communities, creating what Carl Chinn has described as an 'environment of poverty', centred on the slum neighbourhoods or factory quarters of inner cities (1995 [2006], 61–80). The rich, meanwhile, relocated to the safety of the new suburbs, itself a revolutionary development (Luckin 2015, 172). At the same time, environmental impacts were exported, anticipating 'the new ecological imperialism [today] brought about by globalisation' (Faber 2018, 69). Increasingly, and as imperial intervention took the place of less formal if no less exploitative relationships with the developing world, this economic activity assumed a further discriminatory dimension, as assumptions of national, British pre-eminence gave way to racial presumptions of white superiority. Nor was this discriminatory process entirely absent from Britain itself, where poverty was sometimes regarded not only as the fault of the poor, but as 'almost hereditary' (Chinn 1995 [2006], 4; see also Hilton 2006, 581); 'racial analogies' were 'ubiquitous in nineteenth-century social reportage' (Fraser and Brown 1997, 92), sharpening resentment into resistance as a supposedly feckless, inferior underclass redefined itself as the working class. Theirs was also (to use the term used by Guha and Martinez-Alier) an 'environmentalism of the poor' (1997, 18).

Anxiously, Victorians debated what might come of this 'an age of transition', as John Stuart Mill described it in 1831 (228). 'The Old has passed away', wrote Thomas Carlyle in the same year, 'but alas, the New appears not in its stead; the Time is still in pangs of travail with the New' (1831 [1897], Vol. 15, 237). There was, in other words, a self-conscious awareness that this was a time of change – 'the nineteenth century will be known to posterity', wrote Mill, 'as the era of one of the greatest revolutions of which history has preserved the remembrance' (1831 [1986], 228) – but there was no agreement about whether that change would bring good or ill. Instead, the question was continuously discussed by critics and commentators who felt obliged to engage with (rather than retreat from) 'this boundless hubbub' (Carlyle 1831 [1897], Vol. 15, 238). That sense of obligation carried over into the literature of the period, which reflected these concerns in a continuing stream of publications that set out to represent but also debate the direction that society was now taking. Some writers saw that direction as progress; many pointed to its costs as well as its benefits; a handful suggested that society was taking entirely the wrong direction, and argued that, at this critical moment in human history when the ascendancy of industrial modernity was not yet assured, revolution was both necessary and possible, even inevitable.

As these discussions underlined, an often radical response to the fundamental basis of Victorian socioeconomic realities encompassed both a critical reaction to environmental inequality – an inequality made manifest in the spatial segregation of slum and suburb – and a more expansive and constructive sense of what constituted a just society, and to whom or what it should apply. This was not simply a matter of (belatedly) extending consideration to 'Nature', although this too was a factor. Victorian thinkers and writers inherited from their Romantic forebears a Wordsworthian sense of the natural world as a place of rest and refuge distinct from the frenetic, modern, urban environment (Houghton 1957, 79), and a growing recognition that the modern world was eclipsing the little that was left of that 'Nature', prompting tentative efforts towards its conservation. Influential as Romanticism was, however, the Victorians were themselves ambivalent about its emphasis on an apotheosised 'Nature', distinct from human society (Parham 2011, 5). Increasingly, the Victorians recognised their society's unavoidable imbrication in (and impact on) an agential reality, a reality from which 'Nature' could not be abstracted. In particular, Victorian scientific discoveries in the fields of thermodynamics and evolutionary biology emphasised human entanglement with nonhuman and more-than-human worlds (Parham 2017 [2019], 118), discoveries that problematised and threatened to collapse the dualistic divide between (human) culture and (nonhuman) nature. These discoveries also eroded an earlier scientific legacy of thinking about nature as passive, inert, and pliable, and as somehow apart from and not a part of human nature. Consequently, argues John Parham, the Victorians developed a 'broadly materialist awareness that "human being", as moulded by social and political institutions, ultimately resides in the nature and quality of humanity's relationship with other species and its surrounding environment' (2011, 5). As the Victorians increasingly recognised, that environment was no less important – and agential – for being built of bricks and mortar, or for its apparent abstraction from the natural world: the very absence of green spaces was felt to be a problem. Yet a nonhuman nature was still present, even when (as in the cities) it seemed to have been almost entirely excluded. Now, however, it took the distorted form of smoke-filled skies, polluted rivers and streams, and frightening new diseases, like cholera; to critics and commentators, it seemed that the very identity of 'Nature' was changing.

As the Victorians were forced to recognise, observe Laurence W. Mazzeno and Ronald D. Morrison, urban environments could also be nightmarish, toxic environments (2019, 5). The Victorians were encountering, in other words, the advent of what Ulrich Beck has described as a risk society, a 'self-endangering civilization' (Beck 1992 [2012], 10) in which new risks lay bare human vulnerabilities and what Plumwood describes as 'the reality of our embeddedness', 'of ourselves as ecologically constrained beings' (2002, 97).

As Beck has suggested, industrial society automatically generates new forms of risk, forcing society into a radical reconsideration of the terms of its own modernisation. But as Beck also argued, risk may not (and may never) be explicable or comprehensible (Beck 1994, 6), creating a broader, cultural crisis, with 'epochal' (1994, 7) consequences for the way in which industrial society is regarded, not as a stable, progressive domain, but as one whose foundational structures are being called into question. In all its forms, matter and a material world appeared to the Victorians in a new guise, as active rather than passive, and capable not only of registering human impacts, but of reflecting them back onto Victorian society. Thus, and whilst the Victorians were very far from anticipating every aspect of new materialist thinking – thinking that is, in the case of Barad's agential realism, derived from twentieth-century quantum physics – they were similarly aware of reality as itself unpredictable and agential, of their own inextricable involvement with it, and of the potentially dangerous effect of industrial modernity and the discursive structures that underpinned it. The risks and hazards that were being created, as they sensed, would rebound, and most often on those least responsible for and least able to cope with them (Parham 2017 [2019], 116).

In these tentative new ways of thinking about the world and their place in it, Victorian writers began to recognise the wider impacts of 'progress' as its negative effects expanded outwards in time and space and across species boundaries, creating iniquities and inequalities that challenged existing conceptions of justice. Witnessing its effects, they too began to wonder whether a just society was one that necessarily extended moral consideration to nonhuman and more-than-human communities. Problematic as that thinking was, at a time when human, social inequalities were already so acute, it overlaps with the implications of new materialist theories, which (as I have suggested) point towards a more expansive form of justice.

Where, then, should we look for these responses to injustice, these traces of new thinking in the early decades of Victoria's reign – and how, in turn, might this inform the shape of this book? As Elaine Freedgood writes, literature was 'the place from which the political, economic, and social domains would be criticised, and these criticisms could and did achieve canonical status' (2003, 9). Of these literary interventions, the polymathic jeremiads of the Victorian sages were perhaps pre eminent. Figures such as Thomas Carlyle and John Ruskin 'saw themselves, and indeed were seen by their contemporaries, as perfectly able to comment on political economy as well as on art, literature, and culture' (Freedgood 2003, 8–9), and works like Carlyle's *Past and Present* (1843) and Ruskin's *Unto this Last* (1860) played a notable part in shaping social opinion. Under pressure from and in reaction to the emergence of industrial modernity, literature also generated new forms. These included, most notably, the industrial novel (Gallagher 1985, xi), which flourished briefly in the early years of Victoria's reign, and encompassed notable works such as Benjamin Disraeli's *Sybil* (1845), Elizabeth Gaskell's *Mary Barton* (1848) and *North and South* (1854–5), and

Charles Dickens's *Hard Times* (1854), all still in print today. Other new literary forms included histories of industry, industries, and industrialists, books about factory processes and abuses, and – for present purposes, most revealing of all – investigations into the 'condition' of the working classes (Freedgood 2003, 9–13). These included James Phillips Kay-Shuttleworth's *The Moral and Physical Condition of the Working Classes* (1832), Peter Gaskell's *Artisans and Machinery: The Moral and Physical Condition of the Manufacturing Population* (1836), and perhaps most notably, Friedrich Engels's *The Condition of the Working Class in England* (1845; 1887 [New York]; 1891 [London]). Less serious forms of printed matter also played their part in society's search for self-understanding. Relatively inexpensive family journals like Dickens's *Household Words* (1850–9) offered readers access to a variety of fiction and non-fiction, from serialised novels such as Dickens's own *Hard Times* and Elizabeth Gaskell's *North and South*, to reportage, reviews, and commentary, often with a focus on the major social issues of the day. Moreover, and as popular movements such as Chartism demonstrated, cheaper and more ephemeral forms of literature such as newspapers and pamphlets also provided those most affected by environmental injustice with a way of sharing their first-hand experiences of the social and ecological impact of industrial modernity.

As this brief overview suggests, literature in all its forms played a conspicuous and important part in society's attempt to understand the social and ecological impact of industrial modernity, particularly during the turbulent, early years of Victoria's reign. The result was an enormous body of work, and since it cannot all be discussed in depth, I have necessarily focused on a selection of the most significant texts from the period. Some are still well known today; others are now neglected; together, they build up a compelling picture of the lived reality of environmental injustice, inequality, and classism. My starting point, and the focus of the first chapter, is the work of Thomas Carlyle. In 'Chartism', his essay of 1839, Carlyle coined the term the 'Condition of England' (Carlyle 1839 [1897], Vol. 16, 204) to denote the growing divide between rich and poor; as he recognised, this divide was embodied in their now markedly different living and working conditions. As I discuss in the second chapter, these ideas were taken up by Friedrich Engels in *The Condition of the Working Class*, whose 'classic' status Buell highlights (2005, 120) and Bell acknowledges, seeing in it an early response to environmental classism (Bell 2020, 2).

Like Carlyle's essay on 'Chartism', *The Condition of the Working Class* is (ostensibly) a non-fictional critique of the state of society, but as I noted earlier, it was a problem to which many middle-class novelists themselves responded. However, works of fiction such as Gaskell's *Mary Barton* were written, in part, to reconcile the division to which Disraeli drew attention in *Sybil; or, The Two Nations*. In so doing, these middle-class narratives revealed their origins in an ameliorative discourse that sought to defuse conflict without offering fundamental solutions to the iniquities that engendered it. For many of these middle-class writers, the real threat to society came from Chartism, the working-class movement that sprang up in response to the profound inequalities created by industrial modernity. As I discuss in the third chapter of the book, Chartism generated its own, now neglected literary response to environmental injustice, which extended to short stories, poems, and (exceptionally) novels, all intended to particularise working-class experiences. In so doing, these Chartist pioneers sought to counter a narrative of environmental determinism that constructed the working classes as 'one great phantasmagoria' of a 'mad, bad, and dangerous people' that (like 'an infectious disease') threatened 'to destroy civilisation' (Hilton 2006, 581).

The work of Carlyle, Engels, and the Chartists was written at a point in Victorian history when social strife – and its concomitant, environmental degradation – was at its most acute. 'In developing capitalist systems', writes the historian Boyd Hilton, 'there is often a point at which inequality and absolute poverty peak before both are reined back. In Britain, the second quarter of the nineteenth century marked that point' (Hilton 2006, 573). As Hilton adds, '[t]he 1830s and 1840s may well have been the worst ever decades for life expectancy since the Black Death in the history of those parishes which were now experiencing industrialisation' (2006, 574). With the collapse of Chartism in the late 1840s, however, protest died away; in the aftermath of the Great Exhibition, in 1851, it felt to many as though society had entered a new age of relative prosperity and social equilibrium. Some disagreed, most notably Charles Dickens, whose novels took on an increasingly dark tone as he struggled with the unreconciled inequalities that 'progress' had created, in spite (or perhaps because) of itself. The next chapters examine two of his most important novels, *Bleak House* (1852–3) and *Hard Times* (1854). The novels usefully highlight what P. J. Keating has identified as separate strands within Victorian depictions of working-class living conditions: the urban and the industrial (1971 [1979], 11). The urban novel, Keating argued, was an extension of an existing literary tradition, which took on a new relevance as rapid urbanisation worsened conditions in the slums of cities such as London (1971 [1979], 11); here, the impact of industrialisation was often felt indirectly. In the industrial novels, however, the collision between industry, capital, and modernity became the main subject, as these interlinked forces reconstituted urban existence, creating 'two large, economically interdependent [but] conflicting class groups' (Keating 1971 [1979], 11).

This distinction between urban and industrial frames my discussion of Dickens's *Bleak House*, which has become something of an *ur-text* for ecocritics, and the novel which he wrote immediately after it, *Hard Times*. If the former is contiguous with the traditional depiction of slums-as-mysteries, the latter engages directly with the emergence – to contemporaries, it seemed almost overnight – of the factory town. Thus, *Bleak House* represents a kind of environmental justice mystery, in which the existence of a London 'rookery', or slum, is traced to the labyrinthine workings of a self-interested legal system, which colludes with capitalism whilst denying effective participation to those implicated in its workings. In a risk society, however, the harms that slums engender cannot be safely contained: disease and perhaps even violence radiate from them, reminders of society's own complicity in their existence. In *Hard Times*, by contrast, that threatening, destabilising process is writ large, in a factory town where environmental injustice has developed into environmental classism, as the 'hands' (a term which accurately captures the way in which the mill-owners see their workers) live out their lives within the polluted, toxified, and crushingly monotonous environs of Coketown. Violence threatens; what other options do the 'hands' have? Riven by class conflict, a now urbanised society appears to be on the brink of collapse.

As writers of the period began to realise, however, it was by no means easy to undo the conjoined processes of industrialisation and urbanisation, still less engineer some kind of return to the land, important as the idea of it might be as the imagined locus of a sustainable way of life beyond reach of industrial modernity: in *Bleak House*, Dickens allows his heroine to create her own community, far from the soot and fogs of London, but it is a pastoral idyll rather than practical alternative. As the Chartist movement discovered, through its own efforts to create self-sustaining rural communities, it was simply not possible to resist capitalism's relentless transformation of society into an acquisitive, cynical, competitive organism, which 'knows the price of everything, and the value of nothing' (the quip is Oscar Wilde's; 1899, 95).

It is this sense of what a society ought to value – honour, cherish, protect – that informs the work of John Ruskin, to which I turn in the book's sixth and final chapter. Of all the Victorian sages, Ruskin was perhaps the most influential, and in some senses, conventional: his societal ideal was both hierarchical and patriarchal. Yet he also became one of society's most insightful and bold critics. What began as Ruskin's commentary on art and architecture soon became a corrosive analysis of the inequalities that industrial modernity fostered, the kind of work it created, and the distorted, unhealthy environment that was its concomitant. Above all, Ruskin offered a profound critique of the values to which a *laissez-faire* society aspired: its worship of Mammon, and 'the "Goddess of Getting-on"' (*The Crown of Wild Olive*, 1866; Cook and Wedderburn 1903–12, Vol. XVIII, 448).

Ruskin's critique culminated in the four essays of *Unto this Last*, first published in 1860. 'THERE IS NO WEALTH BUT LIFE', he famously declared (Cook and Wedderburn 1903–12, Vol. XVII, 105), and as I discuss, Ruskin's strong sense of what that life ought to entail points towards the kind of inclusive, emergent justice with which this book is, ultimately, concerned. Industrial modernity, wrote Ruskin, spawned 'diminished lives in the midst of noise, of darkness, of deadly exhalation', lives that were markedly worse for the working classes; a flourishing existence required 'pure air [and] pure water' (Cook and Wedderburn 1903–12, Vol. XVII, 110); and '[a]s the art of life is learned, it will be found at last that all lovely things are also necessary; – the wild flower by the wayside, as well as the tended corn; and the wild birds and creatures of the forest' (Vol. XVII, 111). All these things were necessary, Ruskin insisted, in a world in which the material and the discursive are intimately entangled, in 'every wondrous word and unknowable work of God' (Cook and Wedderburn 1903–12, Vol. XVII, 111).

For Ruskin, however, the publication of his essays marked an end as well as beginning. Although he would elaborate on the ideas (both sociocentric and ecocentric) that inform *Unto this Last*, and himself advocate the need for rural (but feudal) communities (Hardy 1979, 78–9), his views made him an increasingly marginal figure. For many in society, the Condition-of-England Question had been answered. Amidst growing prosperity, society settled into years of relative economic and political quiescence. The crisis had passed, conflict had been averted, and after decades of heated literary intervention in the social and ecological problems of the day, an interregnum opened up in the literature of the period. But the lived reality of environmental injustice had not gone away; if anything, it intensified, in those corners of town and city into which society preferred not to look too closely, if at all. Later in the century, society would (re)discover that lived reality, with disturbing implications for what it saw as the trajectory of modern society. As I explain in the concluding chapter, the issue was not simply the impact of environmental classism, the spatial and psychological segregation of slum and suburb, and the growing resentments to which this visible form of inequality gave rise. It was also a sense of what Derek Gladwin has called ecological exile (2018, 5), which extended even to those whose lives seemed outwardly affluent. Something was lacking, and that lack was a constitutive engagement with the (itself threatened) nonhuman and natural world. What was society to do, and how was it to do it? These questions were taken up by a later generation of Victorian thinkers and writers, including Matthew Arnold and William Morris, and in new forms of literature, such as slum fiction and speculative fiction. In turn, these works look forward to our own, late-modern predicament, and the continuing realities of environmental inequality and environmental classism.

This, then, is the outline of the book, which explores the Victorian experience of environmental injustice through its own literature. Before turning to that discussion, however, I want to glance back to the dusty wasteland – the nuisance neighbourhood – with which the Introduction opened, and ask: what might we now make of the grim, sepulchre world on which *Household Words* reported in 'Dust; or ugliness redeemed' (1850, 379–84)? Ecojustice materialism focuses on the interplay between the material and social/discursive world, and its relationship to questions of (in)justice. Drawing on Barad's theory of an agential reality, this interplay is properly understood as an 'intra-action' (2007, 33), from which separable entities emerge; in this co-constitutive process of becoming, matter itself plays a dynamic role in constituting social relations, discursive regimes, and concepts of who and what should be included in discussions of environmental justice; who or what, that is, deserves recognition and participation, and what capabilities society should promote as integral to its own flourishing.

This sense of matter's dynamism – of reality as agential – emerges from *Household Words*'s own article. Here, the dust-mounds stand as paradoxical proof not of the entropic and unsustainable nature of a fossil-fuel-powered economy, but of matter's transformative potential, as it awaits its own recycling (Mershon 2016, 484) – whether into bricks, roads, or paper ('Dust' 1850, 380). This is not a wasteland, made up of dead matter, but of things continuing 'to *be*' (Simon 2011, 219), thereby demonstrating what Ella Mershon calls their 'absolute fungibility', their power to 'circulate perpetually' (2016, 482). But there is a disturbing corollary to this story of 'ugliness, redeemed', which so conveniently wards off the troubling connection between life and pollution (MacDuffie 2014, 127), and between this new economic order and the waste it produces on such a vast scale.[12] It requires, as Allen MacDuffie emphasises, 'the additional energy extracted from the bodies of the lower classes' (2014, 129); it requires a social residue to do the waste-picking, that most degrading, dispiriting, and perhaps dangerous of jobs. It requires, in other words, the broken-down bodies of Peg, Gaffer, and Clem, whose infirmities render them unfit for any other kind of work. Yet work they must, if they are to survive.

Here, a new materialist analysis dovetails with a materialist one: under the conditions created by capitalism, Marx wrote in his 'Economic and Philosophical Manuscripts' (1844), 'labour, *life activity, productive life* itself appears … only as a *means* for the satisfaction of a need, the need to preserve physical existence' (italics in the original; Marx 1975 [1992], 328). For Marx, this was a pivotal shift. To lose control over one's own labour – and the products of it – was to suffer a profound estrangement from what Marx described as *Gattungswesen*, a species-being (or, in new materialist terms, species-becoming) that encompasses not just the self in relation to society (for '[t]he individual *is* the *social being*' [italics in the original; Marx 1975 [1992], 350]) but the self as a 'living natural being' (Marx 1975 [1992], 389), whose flourishing is co-dependent on a 'wealth of human-natural interconnections' (Foster and Burkett 2016 [2017], 53).

Marx's own materialist analysis can, in turn, be read back into the predicament of Peg, Gaffer, and Clem, alienated products of a growing capitalist world-economy. Exiled by a society that renders them all but invisible, reduced to picking over its bones, they live in ecological exile (Gladwin 2018, 5) in a world whose fragmentation mirrors their own; here, 'nature' is reduced to the totemic yet somehow uncanny presence of a crow or two and a pig rooting around in the rubbish ('Dust' 1850, 380). Nature finds a way – but who can claim that this amounts to a flourishing or fair existence? It is, after all, a question of (emergent) justice: it follows from Marx's sense of *Gattungswesen* as species-being (or species-becoming) that nonhuman as well as human intra-actions are themselves an integral part of what constitutes a flourishing, and by extension a good life. Within the horizon marked out by the mountainous dust-heaps, however, nonhuman nature exists in a state of near total and 'perpetual withdrawal' (Taylor 2016, 5). It can hardly be otherwise: coal ash constitutes a profoundly toxic environment, as numerous scientific studies now show (Kravchenko and Lyerly 2018), and *Household Words*'s description of the 'great, black' ('Dust' 1850, 380), 'dirty black' (379) dust-mounds itself implies.[13]

For Peg, Gaffer, Clem, and the other waste-pickers (both human and nonhuman) that make up the dust-mound community, theirs is a life defined by environmental injustice, in each of the principal ways in which it is now theorised – whether in terms of distributive or procedural/participatory justice, justice as recognition, or justice as capabilities (Holifield, Chakraborty, and Walker 2018, 4). Environmental harms are unfairly heaped upon the waste-pickers by a society in whose workings they are denied participation; like the pariahs with whom they are compared, they are given no recognition – hence 'nonrecognition' – and deprived of any meaningful way in which to develop the capabilities that might enable them to flourish. Their existence is instead defined by the struggle to survive. Yet that existence is not a localised or temporary aberration; it is rather the function of a systematic form of attrition or 'slow' violence (Nixon 2011, 2) inherent in the smooth functioning of industrial modernity (Žižek 2008 [2009], 1; I take up this point in Chapter 5).

For Peg, Clem, and Gaffer, therefore, and unlike the 'hard-' and 'software' they help to circulate as 'reclaimers' – the modern term, used in preference to 'scavengers' (WIEGO n.d.) – there is no redemptive arc. For the readers of *Household Words*, whom its 'conductor' Dickens conceived of in terms of fireside and family, this was an unsettling conclusion. But what the real world failed to supply – a happy ending – *Household Words* could itself arrange. In Victorian literature, unexpected inheritances were a favourite plot device for solving the problem of poverty. *Household Words* contrives the same result, when Peg, Gaffer, and Clem find paper proof of a man's fortune and are rewarded by him with their heart's desire: 'a cottage in the neighbourhood of the Dust-heap, built large enough for all three to live together, and keep a cow' ('Dust' 1850, 384). What becomes of the other, nameless scavengers who pick over the dust-mounds is left unrecorded. The reader's eye is instead directed to these improbably happy few, who are lifted (briefly) out of their obscurity, before being returned to it – still safely hidden away behind a mountain range of refuse.

In 'Dust; or Ugliness Redeemed', what we have, in other words, is a complex example of the interaction between material realities and discursive construction: between the manifest environmental injustice of the dust-mounds, a degraded and degrading world in which society's refuse is heaped and its cast-off citizens immured, and its transformation into a narrative that confirms rather than contradicts a middle-class readership's faith in that shibboleth of modernity, 'progress'. In turn, and by reducing the story of the dust-mounds and their scavengers to the story of Peg, Jem, and Gaffer, and eliding their 'bodily experience of scavenging' (MacDuffie 2014, 129), *Household Words* avoids the obvious question: who or what is responsible for their ecological exile, trapped within the mountain range of the dust-mounds that bounds their lives both physically and mentally? The article ignores the wider, structural dimension of environmental injustice, and its relationship to broader questions of environmental inequality and ecological despoliation, just as it is also avoids asking the equally unsettling question: what might come of it all? The mounds were a locus of Victorian anxieties about the unpredictable nature of a world reshaped by industrial modernity: who knew what toxins they harboured, diseases they engendered, or revolutionary sentiments they might encourage? Who knew what might come of this agential reality – this world of animate objects, and of people treated as things? As I discuss in the Conclusion, these were anxieties a later generation of Victorian writers were not afraid to expose.

*

Notes

1 For further background on the area's development in the nineteenth century, see Baggs, Bolton, and Croot (1985, 69–76) and Darley (2018–19, 10–11).

2 For a recent analysis of the living and working conditions of the world's waste-pickers, see Uddin et al. (2020); their analysis underlines the extent to which the conditions encountered in passing by R. Wilfer and described by *Household Words* persist today, 'primarily in low-income and middle-income countries but to a smaller extend also in high-income countries' (2020, 833).

3 Historians often divide Victoria's long reign into three unequal parts, referring to the period 1837–51 as 'early Victorian' (e.g. Harrison 1988). This makes a certain sense, since 1851 marks something of a social watershed, as a decade of hunger and strife gave way to a period of relative prosperity and calm (Harrison 1988, 11–12). Yet social inequality did not simply disappear, and nor did the literature of dissent to which it gave rise: two of Charles Dickens' most substantial interventions in the debate, *Bleak House* and *Hard Times*, date from 1852–3 and 1854, whilst it was not until the 1850s that art critic John Ruskin began that turn towards social criticism that develops into *Unto this Last*, published as essays in 1860. Nor did Chartism and Chartist literature simply wither away in 1848. With these points in mind, I have chosen to focus on the period 1837–60.

4 It follows, Moore argues, that green thinking's conventional emphasis on 'the Industrial Revolution as *the* turning point in human affairs' (2017, 3) occludes 'capitalism's early-modern origins, and its extraordinary reshaping of global natures long before the steam engine' (3). It also follows that environmental justice scholarship should look still further back than the nineteenth century. However, not all scholars are convinced by this rephrasing of the Anthropocene as Capitalocene; see Morton (2016, 23).

5 'Seizing upon anything but class', wrote Michael Parenti in 1997, 'U.S. leftists today have developed an array of identity groups centring around ethnic, gender, cultural, and life-style issues. These groups treat their respective grievances as something apart from class struggle, and have almost nothing to say about the increasingly harsh politico-economic class injustices perpetrated against us all' (151). 'The contemporary Left claims not to exist', observes Jodie Dean (2012, 53). 'Whereas the Right sees left-wing threats everywhere, those on the Left eschew any use of the term "we", emphasizing issue politics, identity politics, and their own

fragmentation into a multitude of singularities' (Dean 2012, 53). Dean's assumption is, in part, that collectivist politics presume universalist positions.

6 For a detailed discussion of the historical emergence of this divide in relation to race, gender, and nature, see Nicholson 2008, pp. 9–34; here, Linda Nicholson interrogates the emergence of these social categories from the late eighteenth to the late nineteenth centuries, and the role that science played in surrounding those categories in an 'aura of objectivity of neutrality' (9).

7 Following early but isolated interventions such as Joni Adamson's *American Indian Literature, Environmental Justice, and Ecocriticism* (2001) and Jeffrey Myers' *Converging Stories: Race, Ecology, and Environmental Justice in American Literature* (2005), there is now a growing body of ecocritical scholarship that draws on or references the idea of environmental justice, such as Eóin Flannery's *Ireland and Ecocriticism: Literature, History and Environmental Justice* (2018), Laura A. White's *Ecospectrality: Haunting and Environmental Justice in Contemporary Anglophone Novels* (2020), and Janet Fiskio's *Climate Change, Literature, and Environmental Justice* (2021). What many of these texts arguably lack, nevertheless, is a willingness to engage more thoroughly in the critical apparatus that environmental justice scholarship has developed in the decades since its inception.

8 For a further discussion of recent developments in the field of Victorian ecocriticism, see Mazzeno and Morrison 2019, pp. 1–5, and Tait 2021, pp. 105–6. The field is developing, quickly.

9 Its title notwithstanding, Newman's monograph is concerned with what he describes as 'the Romantic [i.e. nineteenth] century, when capitalism and environmentalism together spread around the Atlantic rim and across the planet' (2019, 8).

10 Such an acknowledgement is, in part, consistent with the realist position advocated by first-wave ecocritics (Parham 2010, 48; Soper 1995, 130), since it too implies that the human body (like nature itself) is not an exclusively cultural construction, but one with its own 'natural dimension' (Soper 1995, 129), its own 'biological dependency on the eco-system' (Soper 1995, 130). The difficulty of this position, as Kate Soper acknowledges, is the 'ideological use to which' 'invocations of nature and biology' 'have been put' (Soper 1995, 129), not least in the construction of a 'woman-nature equivalence that has served as legitimation for the domestication of women and their relegation to maternal and nurturing functions' (123).

11 Difficult as it might be to conceive of practical solutions to the problems an emergent justice might generate – since it implies equal status under the law for human and nonhuman, but also biotic and abiotic communities, societies, and assemblages or entities – a substantive strand of criminological theory is now focused on 'green justice'. Rob White's *Crimes Against Nature: Environmental Criminology and Ecological Justice* (2008) was an early, ground-breaking intervention in the discussion, along with *Issues in Green Criminology: confronting harms against environments, humanity and other animals* (2007), the collection of essays edited by Piers Beirne and Nigel South. More recent surveys include Rob White and Diane Heckenberg, *Green Criminology: An Introduction to the Study of Environmental Harm* (2014) and Angus Nurse, *An Introduction to Green Criminology and Environmental Justice* (2016).

12 By the time *Our Mutual Friend* was published in volume form, the sustainability of such an energy-intensive economy was already in question; W. Stanley Jevons' *The Coal Question; An Inquiry Concerning the Progress of the Nation, and the Probable Exhaustion of Our Coal Mines*, published in 1865, highlighted the nation's astonishing increase in coal use. When *Household Words* published its article in 1850, London was importing 3,638,883 tons of coal a year, an increase

of 231% over the previous 50 years; just 15 years later, Jevons recorded, the figure had increased to 5,909,940 tons (Jevons 1865, 50).

13 Julia Kravchenko and H. Kim Lyerly (2018) synthesise 30 years of research into the impact of coal power and the coal-dust (or fly-ash) it generates, highlighting the daunting number of ways in which the use of coal affects human health both in the near and long term; whilst their research does not extend to nonhuman impacts, it is reasonable to assume that 'the cumulative impacts of multiple air, water, and soil contaminants' would also affect other biotic communities (Kravchenko and Lyerly, 2018, 289). Strictly, the dust-mounds described by *Household Words* contain other forms of refuse, up to and including human excrement (Brattin 2002, 27), but coal or fly-ash remained the main component, hence the waste-pickers' complaints when it flies up into their faces as they sift the mounds ('Dust' 1850, 380). Note that these 'cinders and ashes' ('Dust' 1850, 380) were not necessarily a product of industry; everywhere in Victorian Britain, homes relied on open coal-fires to keep warm, heat water, and cook. Today, coal remains 'the largest source of energy on earth', extensively used in industry (Munawer 2018, 87).

Works cited

Adamson, Joni. 2001. *American Indian Literature, Environmental Justice, and Ecocriticism*. Tucson, AZ: University of Arizona Press.

Adamson, Joni, Mei Mei Evans, and Rachel Stein. 2002. 'Introduction: Environmental Justice Politics, Poetics, and Pedagogy'. *The Environmental Justice Reader*, edited by Joni Adamson, Mei Mei Evans, and Rachel Stein, 3–14. Tucson, AZ: University of Arizona Press.

Alaimo, Stacey. 2010. *Bodily Natures: Science, Environment, and the Material Self*. Bloomington, IN: Indiana University Press.

Alaimo, Stacey. 2016. *Exposed: Environmental Politics and Pleasures in Posthuman Times*. London: University of Minnesota Press.

Alaimo, Stacey, and Susan Hekman. 2008. 'Introduction: Emerging Models of Materiality in Feminist Theory'. *Material Feminisms*, edited by Stacey Alaimo and Susan Hekman, 1–19. Bloomington, IN: Indiana University Press.

Baggs, A. P., Diane K. Bolton, and Patricia E. C. Croot. 1985. 'Islington: Economic history'. *A History of the County 1997 of Middlesex: Volume 8, Islington and Stoke Newington Parishes*, edited by T. F. T. Baker and C. R. Elrington, 69–76. London: Victoria County History. Available at British History Online, www.british-history.ac.uk/vch/middx/vol8/pp69-76. Accessed 19 October 2022.

Barad, Karen. 2007. *Meeting the Universe Halfway: Quantum Physics and the Entanglement of Matter and Meaning*. London: Duke University Press.

Barad, Karen. 2012. 'Interview with Karen Barad: 'Matter Feels, Converses, Suffers, Desires, Yearns and Remembers''. *New Materialism: Interviews & Cartographies*, edited by Rick Dolphijn and Iris van der Tuin, 48–70. Ann Arbor, MI: Open Humanities Press.

Beck, Ulrich. 1992 [2012]. *Risk Society: Towards a New Modernity*, translated by Mark Ritter. London: Sage.

Beck, Ulrich. 1994. 'The Reinvention of Politics: Towards a Theory of Reflexive Modernization'. *Reflexive Modernization: Politics, Tradition and Aesthetics in the Modern Social Order*, edited by Ulrich Beck, Anthony Giddens and Scott Lash, 1–55. Cambridge: Polity.

Beck, Ulrich. 1995 [2002]. *Ecological Politics in an Age of Risk*, translated by Amos Weisz. Cambridge: Polity Press.
Beirne, Piers, and Nigel South, eds. 2007. *Issues in Green Criminology: Confronting Harms Against Environments, Humanity and Other Animals*. Devon: Willan.
Bell, Derek, and Jayne Carrick. 2018. 'Procedural Environmental Justice'. *The Routledge Handbook of Environmental Justice*, edited by Ryan Holifield, Jayajit Chakraborty, and Gordon Walker, 101–12. Abingdon: Routledge.
Bell, Karen. 2014. *Achieving Environmental Justice: A Cross-National Analysis*. Bristol: Policy Press.
Bell, Karen. 2020. *Working-Class Environmentalism: An Agenda for a Just and Fair Transition to Sustainability*. Cham, Switzerland: Palgrave Macmillan.
Bennett, Jane. 2010. *Vibrant Matter: A Political Ecology of Things*. London: Duke University Press.
Boone, Christopher G., and Geoffrey L. Buckley. 2018. 'Historical approaches to environmental justice'. *The Routledge Handbook of Environmental Justice*, edited by Ryan Holifield, Jayajit Chakraborty, and Gordon Walker, 222–30. Abingdon: Routledge.
Brattin, Joel J. 2002. 'Constancy, Change, and the Dust Mounds of *Our Mutual Friend*'. *Dickens Quarterly* 19 (1): 23–30. Available at jstor.org, www.jstor.org/stable/45291848. Accessed 16 October 2022.
Buell, Lawrence. 2001. *Writing for an Endangered World: Literature, Culture, and Environment in the U.S. and Beyond*. London: Belknapp Press.
Buell, Lawrence. 2005. *The Future of Environmental Criticism: Environmental Crisis and Literary Imagination*. Oxford: Blackwell.
Bullard, Robert D., Glenn S. Johnson, and Beverly H. Wright. 1997. 'Confronting Environmental Injustice: It's The Right Thing to Do'. *Race, Gender & Class* 5 (1): 63–79. Available at jstor.org, www.jstor.org/stable/41674849. Accessed 5 January 2023.
Bullard, Robert D., Glen S. Johnson, and Angel O. Torres. 2005. 'Addressing Global Poverty, Pollution, and Human Rights'. *The Quest for Environmental Justice: Human Rights and the Politics of Pollution*, edited by Robert D. Bullard, 279–97. Berkeley, CA: Counterpoint.
Carlyle, Thomas. 1831 [1897]. 'Characteristics'. *The Works of Thomas Carlyle, Volume Fifteen*, 208–47. New York, NY: Peter Fenelon Collier. Available at archive.org, https://archive.org/details/carlylecompletew15carl/page/n7/mode/2up?view=theater. Accessed 24 December 2022.
Carlyle, Thomas. 1839 [1897]. *Chartism. The Works of Thomas Carlyle, Volume Sixteen*, 204–85. New York, NY. Peter Fenelon Collier. Available at archive.org, https://archive.org/details/worksofthomascar0016carl/page/n7/mode/2up?view=theater. Accessed 24 December 2022.
Chinn, Carl. 1995 [2006]. *Poverty amidst Prosperity: The Urban Poor in England, 1834–1914*. Lancaster: Carnegie.
Clark, Timothy. 2011. *The Cambridge Introduction to Literature and the Environment*. Cambridge: Cambridge University Press.
Cook, E. T. and Alexander Wedderburn (eds). 1903–12. *The Works of John Ruskin*, Library Edition. 39 vols. London: George Allen. Available at www.lancaster.ac.uk/the-ruskin/the-complete-works-of-ruskin/. Accessed 5 January 2023.
Crutzen, Paul, and Eugene F. Stoermer. 2000. 'The 'Anthropocene''. *Global Change Newsletter* 41: 17–18. doi:10.1007/978-3-030-82202-6_2

Darley, Peter. 2018–19. 'The Copenhagen Tunnel'. *Journal of the Islington Archaeology & History Society* 8 (4): 10–11.

Dean, Jodie. 2012. *The Communist Horizon*. London: Verso.

Derrida, Jacques. 1992. 'Force of Law: 'The Mystical Foundation of Authority''. *Deconstruction and the Possibility of Justice*, edited by Drucilla Cornell, Michel Rosenfeld, and David Gray Carlson, 3–67. Abingdon: Routledge.

Dickens, Charles. 1839 [1995]. *Sketches by Boz*, edited by Dennis Walder. London: Penguin.

Dickens, Charles. 1864–5 [1997]. *Our Mutual Friend*, edited and introduced by Adrian Poole. London: Penguin.

'Dust; or Ugliness Redeemed'. 1850. *Household Words, Volume 1*, 13 July: 379–84. Available at Dickens Journals Online, www.djo.org.uk/household-words/volume-i/page-379.html. Accessed 6 April 2022.

Eagleton, Terry. 2016. *Materialism*. London: Yale University Press.

Eckersley, Robyn. 2004. *The Green State: Rethinking Democracy and Sovereignty*. London: MIT Press.

Faber, Daniel. 2018. 'The Political Economy of Environmental Injustice'. *The Routledge Handbook of Environmental Justice*, edited by Ryan Holifield, Jayajit Chakraborty, and Gordon Walker, 61–73. Abingdon: Routledge.

Flannery, Eóin. 2018. *Ireland and Ecocriticism: Literature, History and Environmental Justice*. London: Routledge.

Fiskio, Janet. 2021. *Climate Change, Literature, and Environmental Justice: Poetics of Dissent and Repair*. Cambridge: Cambridge University Press.

Foster, John Bellamy, and Paul Burkett. 2016 [2017]. *Marx and the Earth: An Anti-Critique*. Chicago, IL: Haymarket.

Fraser, Hilary, with Daniel Brown. 1997. *English Prose of the Nineteenth Century*. London: Longman.

Fraser, Nancy. 1997. *Justice Interruptus: Critical Reflections on the 'Postsocialist' Condition*. London: Routledge.

Freedgood, Elaine. 2003. 'Introduction'. *Factory Production in Nineteenth-Century Britain*, edited by Elaine Freedgood, 1–18. Oxford: Oxford University Press.

Gallagher, Catherine. 1985. *The Industrial Reformation of English Fiction: Social Discourse and Narrative Form, 1832–1867*. Chicago, IL: University of Chicago Press.

Garrard, Greg. 2012. *Ecocriticism*. 2nd edn. Abingdon: Routledge.

Gladwin, Derek. 2018. *Ecological Exile: Spatial Injustice and Environmental Humanities*. London: Routledge.

Glotfelty, Cheryll. 1996. 'Introduction: Literary Studies in an Age of Environmental Crisis'. *The Ecocriticism Reader: Landmarks in Literary Ecology*, edited by Cheryll Glotfelty and Harold Fromm, xv–xxxvii. Athens, GA: University of Georgia Press.

Godwin, George. 1972. *Town Swamps and Social Bridges*, introduced by Anthony D. King. Leicester: Leicester University Press. First published in volume form in 1859.

Gould, Bryan. 2013. *Myths, Politicians and Money: The Truth behind the Free Market*. Houndmills: Palgrave Macmillan.

Greenwood, James. 1873. *In Strange Company: Being the Experiences of a Roving Correspondent*. London: Henry S. King & Co. Available at archive.org, https://archive.org/details/instarangecompa00greegoog/page/n6/mode/2up. Accessed 3 November 2021.

Guha, Ramachandra, and Joan Martinez-Alier. 1997. *Varieties of Environmentalism: Essays North and South*. London: Earthscan.
Hall, Dewey, W., ed. 2017. *Victorian Ecocriticism: The Politics of Place and Early Environmental Justice*. London: Lexington.
Hardy, Dennis. 1979. *Alternative Communities in Nineteenth Century England*. London: Longman.
Harrison, J. F. C. 1988. *Early Victorian Britain, 1832–51*. London: Fontana Press.
Harvey, David. 2005 [2007]. *A Brief History of Neoliberalism*. Oxford: Oxford University Press.
Hilton, Boyd. 2006. *A Mad, Bad, and Dangerous People? England 1783–1846*. Oxford: Clarendon Press.
Holifield, Ryan, Jayajit Chakraborty, and Gordon Walker. 2018. 'Introduction: The worlds of environmental justice'. *The Routledge Handbook of Environmental Justice*, edited by Ryan Holifield, Jayajit Chakraborty, and Gordon Walker, 1–12. Abingdon: Routledge.
Hoppen, K. Theodore. 1998. *The Mid-Victorian Generation*: 1846–1886. Oxford: Clarendon Press.
Houghton, Walter E. 1957. *The Victorian Frame of Mind, 1830–1870*. London: Yale University Press.
Iovino, Serenella, and Serpil Oppermann. 2014. 'Introduction: Stories Come to Matter'. *Material Ecocriticism*, edited by Serenella Iovino and Serpil Oppermann, 1–17. Bloomington, IN: Indiana University Press.
Jevons, W. Stanley, 1865. *The Coal Question; An Inquiry Concerning the Progress of the Nation, and the Probable Exhaustion of Our Coal Mines*. London: Macmillan. Available at archive.org, https://archive.org/details/in.ernet.dli.2015.224624/page/n1/mode/2up?view=theater. Accessed 23 October 2022.
Keating, P. J. 1971 [1979]. *The Working Classes in Victorian Fiction*. London: Routledge & Kegan Paul.
Kravchenko, Julia, and H. Kim Lyerly. 2018. 'The Impact of Coal-Powered Electrical Plants and Coal Ash Impoundments on the Health of Residential Communities'. *North Carolina Medical Journal* 79 (5): 289–300. doi: 10.18043/ncm.79.5.289
Luckin, Bill. 2015. *Death and Survival in Urban Britain: Disease, Pollution and Environment* 1800–1950. London: Bloomsbury Academic.
MacDuffie, Allen. 2014. *Victorian Literature, Energy, and the Ecological Imagination*. Cambridge: Cambridge University Press.
Marland, Pippa. 2013. 'Ecocriticism'. *Literature Compass* 10 (11): 846–68. doi: 10.1111/lic3.12105
Marx, Karl. 1975 [1992]. 'Economic and Philosophical Manuscripts (1844)'. *Early Writings*, translated by Rodney Livingstone and Gregor Benton, 279–400. London: Penguin.
Mazzeno, Laurence W., and Ronald D. Morrison. 2017 [2019]. 'Introduction: Practical Ecocriticism and the Victorian Text'. *Victorian Writers and the Environment: Ecocritical Perspectives*, edited by Laurence W. Mazzeno and Ronald D. Morrison, 1–12. Abingdon: Routledge.
Mazzeno, Laurence W., and Ronald D. Morrison. 2019. 'Introduction: Representing Victorian Environmental Nightmares'. *Victorian Environmental Nightmares*, edited by Laurence W. Mazzeno and Ronald D. Morrison, 1–21. Cham, Switzerland: Palgrave Macmillan.

Mershon, Ella. 2016. 'Ruskin's Dust'. *Victorian Studies* 58 (3): 464–92. Available at: jstor.org, www.jstor.org/stable/10.2979/victorianstudies.58.3.03. Accessed 16 October 2022.

Mill, John Stuart. 1831 [1986]. 'Spirit of the Age, I'. *The Examiner*, 9 January. *Newspaper Writings by John Stuart Mill, December 1822–July 1831 (Collected Works of John Stuart Mill, Volume XXII)*, edited by Ann P. Robson and John M. Robson, 227–34. Toronto: University of Toronto Press.

Moore, Jason W. 2017. 'The Capitalocene, Part I: On the Nature and Origins of our Ecological Crisis'. *The Journal of Peasant Studies* 44 (3): 1–37. doi: 10.1080/03066150.2016.1235036

Morton, Timothy. 2014. 'Victorian Hyperobjects'. *Nineteenth-Century Contexts* 36 (5): 489–500. doi: 10.1080/08905495.2014.974940

Morton, Timothy. 2016. *Dark Ecology: For a Logic of Future Coexistence*. Chichester: Columbia University Press.

Munawer, Muhammad Ehsan. 2018. 'Human Health and Environmental Impacts of Coal Combustion and Post-Combustion Wastes'. *Journal of Sustainable Mining* 17 (2): 87–96. doi: 10.46873/2300-3960.1125

Myers, Jeffrey. 2005. *Converging Stories: Race, Ecology, and Environmental Justice in American Literature*. Athens, GA: University of Georgia Press.

Newman, Lance. 2019. *The Literary Heritage of the Environmental Justice Movement: Landscapes of Revolution in Transatlantic Romanticism*. Cham, Switzerland: Palgrave Macmillan.

Nicholson, Linda. 2008. *Identity Before Identity Politics*. Cambridge: Cambridge University Press.

Nixon, Rob. 2011. *Slow Violence and the Environmentalism of the Poor*. Cambridge, MA: Harvard University Press.

Nurse, Angus. 2016. *An Introduction to Green Criminology and Environmental Justice*. London: Sage.

Oppermann, Serpil. 2013. 'Feminist Ecocriticism: A Posthumanist Direction in Ecocritical Trajectory'. *International Perspectives in Feminist Ecocriticism*, edited by Greta Gaard, Simon C. Estok, and Serpil Oppermann, 19–36. Abingdon: Routledge.

Parenti, Michael. 1997. *Blackshirts and Reds: Rational Fascism and the Overthrow of Communism*. San Francisco, CA: City Lights Books.

Parham, John. 2010. *Green Man Hopkins: Poetry and the Victorian Ecological Imagination*. Amsterdam: Rodopi.

Parham, John. 2011. 'Editorial: Victorian Ecology'. *Green Letters: Studies in Ecocriticism* 14: 5–9. doi:10.1080/14688417.2011.10589077

Parham John., 2017 [2019]. 'Bleak Intra-Actions: Dickens, Turbulence, Material Ecology'. *Victorian Writers and the Environment: Ecocritical Perspectives*, edited by Laurence W. Mazzeno and Ronald D. Morrison, 114–29. Abingdon: Routledge.

Pellow, David N. 2000. 'Environmental Inequality Formation: Towards a Theory of Environmental Injustice'. *American Behavioural Scientist* 43 (4): 581–601. doi:10.1177/0002764200043004003

Pellow, David Naguib. 2018. *What is Critical Environmental Justice?* Cambridge: Polity.

Plumwood, Val. 1993 [1997]. *Feminism and the Mastery of Nature*. London: Routledge.

Plumwood, Val. 2002. *Environmental Culture: The Ecological Crisis of Reason*. London: Routledge.

Pulido, Laura. 2018. 'Historicizing the Personal and the Political: Evolving Racial Formations and the Environmental Justice Movement'. *The Routledge Handbook of Environmental Justice*, edited by Ryan Holifield, Jayajit Chakraborty, and Gordon Walker, 15–24. Abingdon: Routledge.

Reed, T. V. 2002. 'Towards an Environmental Justice Ecocriticism'. *The Environmental Justice Reader*, edited by Joni Adamson, Mei Mei Evans, and Rachel Stein, 145–62. Tucson: University of Arizona Press.

Schlosberg, David. 2007. *Defining Environmental Justice: Theories, Movements, and Nature*. Oxford: Oxford University Press.

Schlosberg, David. 2013. 'Theorising Environmental Justice: The Expanding Sphere of a Discourse'. *Environmental Politics: Coming of Age? Environmental Politics* at 21 22 (1): 37–55. doi:10.1080/09644016.2013.755387

Seaman, L. C. B. 1973 [1993]. *Victorian England: Aspects of English and Imperial History*, 1837–1901. London: Routledge.

Simon, Leslie. 2011. '*Bleak House*, *Our Mutual Friend*, and the Aesthetics of Dust'. *Dickens Studies Annual* 42: 217–36. Available at jstor.org, www.jstor.org/stable/44371469. Accessed 16 October 2022.

Soper, Kate. 1995. *What is Nature?* Oxford: Blackwell.

Sze, Julie. 2002. 'From Environmental Justice Literature to the Literature of Environmental Justice'. *The Environmental Justice Reader*, edited by Joni Adamson, Mei Mei Evans, and Rachel Stein, 163–80. Tucson, AZ: University of Arizona Press.

Tait, Adrian. 2021 'Eco-criticism(s) and Victorian Fiction: A Review Essay'. *Dickens Studies Annual: Essays on Victorian Fiction* 52 (1): 103–37. doi:10.5325/dickstudannu.52.1.0103

Taylor, Jesse Oak. 2015. 'Where is Victorian Ecocriticism?' *Victorian Literature and Culture* 43 (4): 877–94. doi:10.1017/S1060150315000315.

Taylor, Jesse Oak. 2016. *The Sky of Our Manufacture: The London Fog in British Fiction from Dickens to Woolf*. London: University of Virginia Press.

Thornbury, Walter. 1881. *Old and New London: A Narrative of its History, its People, its Places*. New edn. London: Cassell, Petter, & Galpin. Available at archive.org, https://archive.org/details/in.ernet.dli.2015.102566/page/n9/mode/2up. Accessed 17 November 2021. Originally published in two volumes, 1873–4.

Uddin, Sayed Mohammad Nazim, Jutta Gutberlet, Anahita Ramezani, and Sayed Nasiruddin. 2020. 'Experiencing the Everyday of Waste Pickers: A Sustainable Livelihoods and Health Assessment in Dhaka City, Bangladesh'. *Journal of International Development* 32 (179): 833–53. doi: https://doi.org/10.1002/jid.3479

Weeks, John F. 2014. *Economics of the 1%: How Mainstream Economics Serves the Rich, Obscures Reality and Distorts Policy*. London: Anthem.

White, Laura A. 2020. *Ecospectrality: Haunting and Environmental Justice in Contemporary Anglophone Novels*. London: Bloomsbury Academic.

White, Rob. 2008. *Crimes Against Nature: Environmental Criminology and Ecological Justice*. Devon: Willan.

White, Rob, and Diane Heckenberg. 2014. *Green Criminology: An Introduction to the Study of Environmental Harm*. London: Routledge.

Whyte, Kyle. 2018. 'The Recognition Paradigm of Environmental Injustice'. *The Routledge Handbook of Environmental Justice*, edited by Ryan Holifield, Jayajit Chakraborty, and Gordon Walker, 113–23. Abingdon: Routledge.

WIEGO (Women in Informal Employment: Globalizing and Organizing). n.d. 'Waste Pickers'. Available at wiego.org, www.wiego.org/waste-pickers. Accessed 3 January 2023.

Wilde, Oscar. 1899. *Lady Windermere's Fan: A Play about a Good Woman*. London: Elkin Matthews and John Lane. First performed in 1892.

Žižek, Slavoj. 2008 [2009]. *Violence: Six Sideways Reflections*. London: Profile.

1 Thomas Carlyle's 'Condition-of-England Question'

This chapter explores the work of essayist Thomas Carlyle (1795–1881), one of the most important and provocative thinkers of his age. Two factors shaped Carlyle's thought – German Romanticism, and a deeply felt but unconventional religiosity – and they converged in his social commentaries and critiques. Although now relatively neglected, Carlyle's literary interventions were at the time profoundly influential: they shaped the way that the early Victorians understood a world newly made by industrial modernity, in which '[t]he most striking fact of the day was the misery of industrial proletariat' (Piketty 2014 [2017], 8). As George Eliot remarked in 1855, 'there has hardly been an English book written for the last ten or twelve years that would not have been different if Carlyle had not lived' (Eliot 1971 [2002], 410).

Carlyle was quick to realise that the Victorians were entering (and engineering) a new epoch in human history: 'in very truth it is a "new Era"', he wrote in 1839 ([1897], Vol 16, 253).[1] However, he was also alert to the inequality that was its consequence. '[W]ealth has more and more increased', he observed in 'Signs of the Times' (1829 [1897]), 'and at the same time gathered itself more and more into masses … increasing the distance between the rich and the poor' (Vol 14, 466). This was, moreover, a divide with a spatial dimension. 'Towns were becoming more unhealthy', notes the historian Martin Daunton, 'poisoning themselves with their own human, animal, and industrial wastes' (2000, 63). The rich reacted by moving to the suburbs; the working classes were left to make the best of these poisonous urban environments. Whilst the affluent preferred to ignore the existence of these 'ecological sacrifice zones' (Faber 2018, 62), however, they were 'increasingly obliged to look' (Fraser and Brown 1997, 89). Spurred on by Carlyle's invectives, critics and commentators set out to investigate the iniquitous convergence of poverty, pollution, and slum squalor in Victorian towns, and use their findings to seize the attention of 'their deliberately impercipient social superiors' (Fraser and Brown 1997, 89). By the time Carlyle coined the term in 1839, the 'Condition-of-England Question' (Vol 16) was already a talking point, 'justly regarded [he later wrote, in 1843] as one of the most ominous, and withal one of the strangest, ever seen in this world' (Vol 6, 217). 'Descend where you will into the lower class', he wrote,

DOI: 10.4324/9780367821609-2

> the same sorrowful result discloses itself: you have to admit that the working body of this rich English Nation has sunk or is fast sinking into a state, to which, all sides of it considered, there was literally never any parallel.
>
> (Vol 6, 219)

Given the plight of the working classes, observed Carlyle, the 'Question' was simply this: '[i]s the condition of the English working people wrong; so wrong that rational working men cannot, will not, and even should not rest quiet under it?' (Vol 16, 206). Having established his reputation with a history of the French Revolution, Carlyle was aware of what might come of these turbulent times. He was equally determined to confront the situation: as he wrote in 'Signs of the Times', 'not in turning back, not in resisting, but only in resolutely struggling forward, does our life consist' (Vol 14, 487). If we are, therefore, to understand the widespread emergence of environmental injustice in the early Victorian period – and the sometimes-troubling ways in which critics reacted – we must start with Carlyle, whose own response is as perceptive as it is problematic. To that end, this chapter focuses on three of Carlyle's most notable and influential social commentaries: 'Signs of the Times' (1829 [1897]), *Chartism* (1839 [1897]), and *Past and Present* (1843 [1897]). Together, these works did much to set the terms of the Victorian literary response to environmental inequality and classism.

*

For Carlyle, as for many critics, 'the material degradation of large masses of the population' (Swift 2001, 69) was the 'crying shame of the time' (Tillotson 1978, 67). As Carlyle made clear in his essay on the Chartist movement, his immediate concern lay with 'the condition and disposition of the Working Classes' (Vol 16, 204), and the impact on those classes of what is now defined as environmental classism, a term that relates class to environmental (in) quality (Bell 2020, 3–5). As Karen Bell points out, poverty and race are often discussed in relation to issues of environmental equity, but class is not (2020, 5). Yet class is itself a determining factor in dictating who shoulders 'the environmental burdens for society' (Bell 2020, 3). As with race, note Brendan Coolsaet and Pierre-Yves Néron, so with class: 'cultural hierarchization and disrespect [result in the] unfair distribution of environmental problems and exclusionary decision-making processes' (2021, 59). For marginalised groups to win justice, it follows that they must first win recognition: it is a necessary step towards the fairer distribution of environmental harms and goods, and full participation in the political processes that govern their distribution. Consequently, 'demands for recognition have always been part of the claims of different environmental justice movements' (Coolsaet and Néron 2021, 59). If justice depends on recognition, however, *in*justice is often a function of misrecognition or of nonrecognition: of sections within society that are constructed as marginal or inferior, or whose existence is simply ignored altogether.

In Victorian Britain, the working classes found themselves caught at the intersection of both misrecognition and nonrecognition: their poverty was dismissed as the consequence not of industrial capitalism but of their own indigence or improvidence (hence misrecognition), whilst they themselves were isolated in the slum quarters of the cities, where they could be ignored (nonrecognition). Infamously, the New Poor Law (1834) institutionalised this convergence: poverty was the fault of the poor, and relief from it was constructed as punishment, in workhouses that were little different to prisons; to observers like Pierre-Joseph Proudhon, they were 'simply torture-houses' (1847 [2007], 207). In turn, this denial of 'outdoor relief' to the poor (Carlyle Vol 16, 214) – and their consequent incarceration in the workhouse – enabled the well-to-do to construct the fantasy of a liberal, generous, and above all just society (Charles Dickens would later satirise that fantasy in his characterisation of 'Podsnappery', discussed in the Conclusion). '[I]nstantly as the walls of the workhouse arise', wrote Carlyle,

> misery and necessity fly away, out of sight, – out of being, as is fondly hoped, and dissolve into the inane; industry, frugality, fertility, rise of wages, peace on earth and goodwill towards men do, – in the Poor-Law Commissioners' Reports, – infallibly, rapidly or not so rapidly, to the joy of all parties, supervene.
>
> (Vol 16, 215)

In this way, 'the Victorian upper classes and bourgeoisie … contrived a kind of cultural apartheid which meant that the poor never actually crossed their field of vision or entered their consciousness' (Fraser and Brown 1997, 88). Out of sight, the poor suffered, and in the city's slums, they suffered in conditions created both by their poverty (which forced them to accept whatever accommodation they could find within walking distance of the factories) and by industrialisation itself. 'Is it a green flowery world, with azure everlasting sky stretched over it, the work and government of a God', Carlyle inquired of these working-class neighbourhoods (Vol 16, 228), 'or a murky-simmering Tophet, of copperas-fumes, cotton-fuz, gin-riot, wrath and toil, created by a Demon, governed by a Demon?' The question was rhetorical: for the working (or 'labouring') classes, the reality was a 'black unluminous unheeded *Inferno*' (Vol 16, 229). Carlyle's references to copperas and cotton-fuz (or fuzz) point to the lethal side-effects that accompanied industrial development: copperas is an acutely toxic industrial by-product, and 'fuz' the fibrous dust produced during cotton manufacture, giving rise to a fatal form of lung disease called byssinosis.[2] Such was 'British industrial existence', Carlyle wrote, a 'poison-swamp [and] hideous *living* Golgotha' (italics in the original; 'Latter-Day Pamphlets', Vol 13, 285).

As his description of this 'Dantean Hell' (Vol 16, 228) suggests, Carlyle understood the constitutive importance of 'the material interconnections between the human and the more-than-human world' (Alaimo 2010, 2): in the terms of a new materialist analysis, he understood 'human corporeality as trans-corporeality' (Alaimo 2010, 2), and recognised that what are now characterised as environmental health and environmental justice form the basis of a flourishing, viable, and fair society (3). 'What constitutes the well-being of a man?' asked Carlyle. 'Many things; of which the wages he gets, and the bread he buys with them, are but one preliminary item' (Vol 16, 212). 'Every toiling Manchester, its smoke and soot all burnt', Carlyle wrote in *Past and Present* (Vol 6, 468), 'ought it not, among so many world-wide conquests, to have a hundred acres or so of free greenfield, with trees on it, conquered, for its little children to disport in?' This was, Carlyle insisted, a question of what was right, and environmentally just: a true sense of well-being meant equitable conditions, as the working classes insisted. 'It is "for justice" that [the worker] struggles', wrote Carlyle; 'for "just wages", – not in money alone!' (Vol 16, 219).

> To whatever other griefs the lower classes labour under, this bitterest and sorest grief now superadds itself: the unendurable conviction that they are unfairly dealt with, that their lot in this world is not founded on right, not even on necessity and might, is neither what it should be, nor what it shall be.
>
> (Vol 16, 233)

This injustice – this classism – could not and would not stand:

> [v]ain also is it to think that the misery of one class, of the great universal under class, can be isolated, and kept apart and peculiar, down in that class. By infallible contagion, ... the misery of the lowest spreads upwards and upwards till it reaches the very highest.
>
> (Vol 16, 251)

The root problem, as Carlyle recognised, lay in the structural forces on which this inequality was founded. The first of these forces was an unfettered form of free-market capitalism, curbed only by Corn Laws that conspired to keep bread prices high. This form of capitalism was, Carlyle thundered, a 'serious, most earnest Mammonism grown Midas-eared', which had made 'Cash-payment the one nexus of man to man: Free-trade, Competition, and Devil take the hindmost, our latest Gospel yet preached!' (Vol 6, 379). Such a 'Gospel of Mammonism' (Vol 6, 356) dissolved the very basis of society. 'We call it a Society', he wrote; 'and go about professing openly the totalest separation, isolation' (Vol 6, 357).

> Our life is not a mutual helpfulness; but rather, cloaked under due laws-of-war, named 'fair competition' and so forth, it is a mutual hostility. We

> have profoundly forgotten everywhere that *Cash-payment* is not the sole relation of human beings; we think, nothing doubting, that *it* absolves and liquidates all engagements of man.
>
> (Italics in the original; Vol 6, 357)

Second, there was the ungoverned – it seemed ungovernable – process of rapid industrialisation, and the chaotic urbanisation it drove. Together, industrialisation and urbanisation created the realities of daily life for the working class, in factories where the hours were long, the routine unrelenting, and the work dangerous, and in homes that often amounted to a single room in 'a foul cellar' in the midst of 'dark poison-lanes' (Vol 6, 468). Nor was this the sum of it. Industrialisation created one set of problems; the subtler and more insidious effects of 'Industrialism' (*Sartor Resartus*, Vol 12, 92) another, and it was Carlyle who first gave the phenomenon a name (Swift 2001, 68). 'There is no end to machinery', Carlyle wrote in 'Signs of the Times' (Vol 14, 465). 'Not the external and physical alone is now managed by machinery, but the internal and spiritual also' (Vol 14, 466).

> Men are grown mechanical in head and in heart, as well as in hand … Not for internal perfection, but for external combinations and arrangements, for institutions, constitutions, for Mechanism of one sort or other, do they hope and struggle. Their whole efforts, attachments, opinions, turn on mechanism, and are of a mechanical character.
>
> (Vol 14, 468)

Carlyle's insight is a powerfully suggestive one, and it anticipates the idea of alienation or *Entfremdung* that Karl Marx was soon to develop (*Entfremdung* is further discussed in Chapter 2); this estrangement of the self, argued Marx, is an inevitable product of relations in a capitalist economy, where workers are simultaneously estranged from the products of their labour, from themselves, and from each other, as all alike reduced to the status of objects (Marx 1975 [1992], 328). Perhaps 'Mechanism', wrote Carlyle (Vol 14, 468), was a contributory factor in society's fragmentation, leaving the isolated and estranged individual as but 'atoms' (Vol 6, 469) in an '*Un*nature, [or] what we call Chaos, [that] holds nothing in it but vacuities, devouring gulfs' (Vol 6, 353). There was, moreover, a related concern, linked to Carlyle's sense that 'Philosophy, Science, Art, Literature, all depend on machinery' (Vol 14, 467). In his view, a mechanistic way of thinking had reduced 'the power man has of knowing and believing' to 'the mere power of arranging and communicating'; 'Intellect' was reduced to instrumentalising 'Logic' (Vol 14, 479) in the name of 'adapting means to ends' (465).

This sense of contemporary thought as itself grown 'sceptical, utilitarian, mechanical' (Willey 1949 [1973], 135) overlapped with Carlyle's recognition of a third force at work in a now deeply divided society, a force in which the material and the discursive were profoundly interconnected.

Just as environmentalists and environmental justice scholars today refer to damaging and dualistic ways of thinking as part of the problem of social and environmental impoverishment, so Carlyle recoiled (1831 [1897]) from a way of thinking about 'Nature' as *mere* matter, absent her 'vital action, her manifest purpose and effort' (Vol 15, 210) – absent, that is, the 'radical aliveness' to which new materialism draws attention (Barad 2007, 33). The eighteenth-century, Carlyle argued, had bequeathed the nineteenth a legacy of 'Dryasdust Philosophisms and enlightened Scepticisms ... a memorable *Nightmare-Dream*' (italics in the original; Vol 6, 446) whereby the Universe itself was transformed into 'one huge, dead, immeasurable Steam-engine' (*Sartor Resartus*, Vol 12, 126). Such a conception of it, Carlyle argued, represented a fundamental failure of comprehension, imagination, and above all faith. 'All this haggard epoch, with its ghastly Doctrines, and death's-head Philosophies', he wrote, 'will one day have become ... "the Period of Ignorance"' (Vol 6, 446).

For Carlyle, therefore, the manifestly traumatic forms of environmental inequality that beset his age had deep roots in structural forces, at once material and discursive. Yet solutions were possible – and 'bodies came first' (Tillotson 1978, 69). 'Manchester, with its cotton-fuzz, its smoke and dust, its tumult and contentious squalor, is hideous to thee? Think not so', wrote Carlyle (Vol 16, 263). 'Soot and despair are not the essence of it; they are divisible from it' (Vol 16, 264). Since the worst aspects of industrial squalor *were* 'divisible from it', Carlyle argued, responses could be found, but they did not lie in the hands of the individual. The mantra of self-help was not enough: confronted by 'the conflict between the powers of government and the claims of private interests', Carlyle came out 'without qualification on the side of intervention' (Shelston and Shelston 1990, 66). As Carlyle insisted:

> The Legislature, even as it now is, could order all dingy Manufacturing Towns to cease from their soot and darkness; to let-in the blessed sunlight, the blue of Heaven, and become clear and clean; to burn their coal-smoke, namely, and make flame of it. Baths, free air, a wholesome temperature, ceilings twenty feet high, might be ordained, by Act of Parliament, in all establishments licensed as Mills.
>
> (Vol 6, 468)

The measures for which Carlyle called ('pollution control, improved factory conditions, and the provision of public parks and open spaces in the hearts of the cities') 'were to feature prominently amongst the concerns of Victorian government' (Shelston and Shelston 1990, 67). If, however, the answer lay in an appeal to the state, the problem (as Carlyle acknowledged) was that the state was itself caught up in forms of political and economic thinking that simply served to justify the chaotic inequities that beset the labouring classes. As David Naguib Pellow has pointed out, state power may itself be 'a fundamental obstacle to social and environmental justice' (2018, 22);

it may even be that states are inherently predisposed towards 'exclusionary' and 'anti-ecological' policies (23). Carlyle himself argued that all forms of contemporary governance were entangled with and compromised by the same structural forces (at once material-social and material-discursive) that combined to create the lived realities of environmental injustice: as he wrote, 'all government of the Poor by the Rich has long ago been given over to Supply-and-demand, Laissez-faire and suchlike, and universally declared to be "impossible"' (Vol 6, 359–60). On the one hand, juridical process was broken.

> The man in horsehair wig advances, promising that he will get me 'justice': he takes me into Chancery Law-Courts, into decades, half-centuries of hubbub, of distracted jargon; and does *get* me – disappointment, almost desperation; and one refuge: that of dismissing him and his 'justice' altogether out of my head.
>
> (Carlyle Vol 6, 465)

On the other, parliament was ineffective. '[T]he Gospel of Dilettantism', Carlyle wrote, creates 'a Governing Class who do not govern, nor understand in the least that they are bound or expected to govern' (Vol 6, 360), and produces an 'impotent, insolent Donothingism in Practice and Saynothingism in Speech' (Vol 6, 361). Radicalism had itself failed the people it purported to represent. 'What thing has Radicalism obtained for them', Carlyle demanded to know; 'what other than shadows of things has it so much as asked for them?' (Vol 16, 269). It was inevitable, Carlyle concluded, that the working class would attempt to represent themselves, and inevitably, this would require the kind of changes that the Chartists were then demanding, most notably universal suffrage; as Carlyle acknowledged in his essay on the movement, 'Chartism is one of the most natural phenomena in England' (Vol 16, 271) (a remarkable mass movement, Chartism and the literature it generated is further discussed in Chapter 3).

At this point, however, Carlyle's views diverged – sharply – from Chartism's own progressive agenda. Carlyle's sympathy for the predicament of the working classes was undeniable, but in his view, justice did not lie in political participation or equality before the law (fundamental tenets of what is now thought of as environmental justice). Instead, Carlyle simply reinforced society's misrecognition of the working classes – 'that great dumb toiling class which cannot speak' (Vol 16, 207) – by representing their inferiority as part of the natural order. In the midst of his espousal of their rights to a better life, Carlyle's view of the working classes assumed an essentialist dimension, which took on an explicitly racist edge when, in his essay on *Chartism* (1839 [1897]), he shifted his focus from '[t]he Saxon man', who 'has not sunk from decent manhood to squalid apehood', to the country's population of Irish immigrants, in their 'rags and laughing savagery' (Vol 16, 223). Here, and as Roger Swift explains (2001, 72–3), Carlyle was reiterating the discrimination to which those immigrants were exposed daily, a form of discrimination which itself assumed a dualistic divide (Plumwood 1993 [1997], 43): simply, Carlyle positioned the Irish immigrant as closer to the brutish, the animal, and to unreason – aligning her or him, that is, with an inferiorised 'nature'.

Even as Carlyle acknowledged the injustice of the working-class predicament, therefore, he argued that the solution lay, not in their participation in government, but in better government of them. What was the true meaning of the Chartist's demands?

> Bellowings, inarticulate cries as of a dumb creature in rage and pain; to the ear of wisdom they are inarticulate prayers: 'Guide me, govern me! I am mad, and miserable, and cannot guide myself!' Surely of all 'rights of man', this right of the ignorant man to be guided by the wiser, to be, gently or forcibly, held in the true course by him, is the indisputablest. Nature herself ordains it from the first …
>
> (Vol 16, 241)

Given that the working classes were (in his view) mere 'creatures', dumb and inarticulate, the answer to their problems did not lie in 'Democracy, which means despair of finding any Heroes to govern you' (Vol 6, 422). It lay in '[a]n actual new Sovereignty, Industrial Aristocracy, real not imaginary Aristocracy' (Vol 6, 455), an aristocracy that could be found in those 'Leaders of Industry' who were already 'virtually the Captains of the World' (Vol 6, 474).

Thus, Carlyle first naturalised the inferiority of the working classes, and then reasserted the dominance of a now hegemonic bourgeoisie, apparently oblivious to the inference of his own reasoning – that, as Thomas Piketty points out, 'capitalism automatically generates arbitrary and unsustainable inequalities' (2014 [2017], 1). Carlyle was no less myopic when he came to consider the wider impacts of industrial modernity. In the midst of the very visible environmental problems it was creating, he was spellbound by its transformative potential:

> Prospero evoked the singing of Ariel, and took captive the world with those melodies: the same Prospero can send his Fire-demons panting across all oceans; shooting with the speed of meteors, on cunning highways, from end to end of kingdoms; and make Iron his missionary, preaching its evangel to the brute Primeval Powers, which listen and obey.
>
> (Vol 16, 263)

'Commerce stretches its fibres over the whole earth; sensitive literally, nay quivering in convulsion, to the farthest influences of the earth' (Vol 16, 226). What Carlyle was describing, as Timothy Morton has argued, is the hyperobject that is industrial capitalism, 'massively distributed in time and space relative to humans' (Morton 2013, 1), tantalisingly out of reach or even out of sight, yet all the while 'sucking away at the humans on the level beneath' with a 'vampire-like downward causality' (5) (see also Morton 2010, 130–1; 2016, 8–9). ['C]hanging [its] shape like a very Proteus', as Carlyle put it (Vol 16, 226), industrial capitalism operates often unbidden over vast scales of time and space, restless, rootless, always seeking its own perpetuation: as Karl Marx and Friedrich Engels would observe in *The Communist Manifesto* (1848), industrial capitalism 'must nestle everywhere, settle everywhere, establish connexions everywhere' (2002, 223); and so '[a]ll that is solid melts into air' (223).

For Carlyle, this 'immeasurable Proteus Steam-demon' (Vol 16, 228) was something to be celebrated; through it, the unused potential of the natural world could be realised, and all those unclaimed spaces Western societies had long denoted as 'wasteland' transformed and made productive (Palma 2014, 3). 'We remove mountains, and make seas our smooth highway; nothing can resist us', Carlyle wrote in 'Signs of the Times' (Vol 14, 466). 'We war with rude Nature; and, by our resistless engines, come off always victorious, and loaded with spoils' (Vol 14, 466). To Carlyle, it followed that 'the awakening of a Manchester, on Monday morning, at half-past five by the clock; the rushing off of its thousand mills, like the boom of an Atlantic tide, [was a sound] sublime as a Niagara, or more so' (Vol 16, 263–4).

It followed, Carlyle believed, that the 'condition of England' was not an insoluble problem, nor one requiring revolutionary solutions. With better leaders, reformed governance was possible, and with better governance, a society transformed, guided by the recognition that well-being was not a function of 'plethoric wealth' (Vol 6, 221), but of lives lived well; that, as John Ruskin would later insist (1860) in a conscious echo of Carlyle, 'THERE IS NO WEALTH BUT LIFE ... including all its powers of love, of joy, and of admiration' (Cook and Wedderburn 1903–12, Vol XVII, 105). With that recognition, the forces of industrial modernity could themselves be reformed, purged of the smoke and squalor that was their unfortunate corollary, and freed to continue their extraordinary transformation of the material world.

As John Parham has emphasised, Carlyle's argument illustrates the 'social pragmatism out of which a Victorian ecological critique developed' (2010, 79), at once critical of modernisation, yet itself creatively and constructively engaged with it; his work also highlights the self-evident limits of Carlyle's interpretation of 'emancipatory modernity' (69). But as Parham observes, Carlyle's critique embodies a further, perhaps still more profound tension, between its pragmatic realisation that modernity brought benefits – Carlyle was fond of emphasising that Manchester's cotton mills were, after all, the means by which the many were now cheaply clothed – and the 'romantic sensibility' (Parham 2010, 79) that formed so important a part of Carlyle's idiosyncratic but influential beliefs. It is a tension rooted in Carlyle's sense of the divine at work in the universe, and as I discuss, below, it has significant consequences for his view (and valuation) of the more-than-human world.

Carlyle spoke most powerfully to his contemporaries when he spoke to them of their own religious doubt, and his own spiritual answer to it. It was, to be sure, an unconventional answer, since Carlyle sought meaning in this world – this more-than-human world – rather than the next: he invested the universe with spiritual significance, and found in its dynamism an answer to the mechanistic philosophies of an age that preferred to see that world as dead matter, infinitely malleable and manipulatable. In effect, he argued that intrinsic worth was widely distributed throughout creation, a premise which is itself the basis of a broader understanding of 'environmental' justice – and here, once again, Carlyle's views converge with and anticipate new materialist thinking, in powerfully suggestive ways.

The starting point for Carlyle's response lay in his dismayed sense that the age was becoming a faithless one. 'In our age of Down-pulling and Disbelief', he wrote in *Sartor Resartus* (1833–4 [1897]), 'the very Devil has been pulled down' (Vol 12, 126). In response, Carlyle sought to persuade his readers of the living meaning of what he called the 'EVERLASTING YEA, wherein all contradiction is solved' (*Sartor Resartus*, Vol 12, 146). The urgency of society's contemporary problems simply underlined the need for his readers to grasp this fundamental truth. Significantly, that sense of urgency carries over into Carlyle's unique and unusual writing style. Carlyle is impassioned, argumentative, declamatory; every paragraph rises to and has its moment of drama, and every word or expression is, if not newly made, then used as if for the first time. It is a vital kind of language, and in its message, vitalist: as critics have variously highlighted, Carlyle's style seeks to make 'concrete the vision which it seeks to communicate' (John 1974, 78), and here, that vision relates to a suitably capitalised 'Nature', as the wider world of which humankind is a (tiny) part.[3] '[W]hat is Nature?' asked Carlyle, in a tone reminiscent of a preacher's exhortations.

> Ha! Why do I not name thee God? Art not thou the "Living Garment of God"? O Heavens, is it, in very deed, He, then, that ever speaks through thee; that lives and loves in thee, that lives and loves in me?
>
> (*Sartor Resartus*, Vol 12, 143)

'[I]n the heart of its tumultuous Appearances, Embroilments, and mad Time-vortexes', Carlyle wrote of the universe, 'is there not, silent, eternal, an All-just, an All-beautiful; sole Reality and ultimate controlling Power of the whole?' (Vol 6, 435). 'Sooty Manchester, – it too is built on the infinite Abysses', he insisted; '[g]o or stand, in what time, in what place we will, are there not Immensities, Eternities over us, around us, in us …' (Vol 6, 468).

Carlyle's vitalism could be interpreted, as the theologian F. D. Maurice argued (1840), as a kind of 'wild pantheistic rant' (Maurice 1885, 283), but it was a rant that dissolved another of the interlocking dualisms to which

Plumwood refers, a theological dualism whereby the mind is aligned with spirit, and both are opposed to matter and to nature (Plumwood 1993 [1997], 43). As Carlyle inquired, in 'Characteristics' (1831 [1897]), 'is not that very division of the unity, Man, into a dualism of Soul and Body, itself the symptom of disease; as, perhaps, is your frightful theory of Materialism; of his being but a Body'? (Vol 15, 211). The paradox of Carlyle's position is, of course, that by insisting on the spiritual within the bodily, he was not so much contradicting materialism as elevating it. In spite of the Calvinistic beliefs in which he had been raised, and which play such an important part in shaping his prose, Carlyle's view of religion is (as Basil Willey argued) a 'vigorous continuation' of Romanticism (Willey 1949 [1973], 117), fusing 'supernatural and natural, spirit and matter, sacred and profane' (114). Thus, in 'raising the natural to the supernatural', as Carlyle wrote in a journal entry of 1833 (Froude 1882, 267), his critique does not transcend the material world so much as infuse it with an alternative meaning and value. In the words of Jerome Buckley, Carlyle 'sought the miracle of being', and found it 'in the firm soil of actuality' (1952 [1966], 37).

In spite, therefore, of Carlyle's opposition to 'black materialisms' (*John Sterling*, Vol 13, 52) and mechanistic philosophies – and to the reductionistic rationalism that underpinned them – his highly personal faith itself constitutes what we would now understand as a 'positively material theology' (Keller and Rubenstein 2017, 5). As such, it is both compatible with sometimes overlooked or submerged aspects of Christian thinking (Keller 2017, 111–4) and new materialist discussions (and tentative elaborations) of a material world charged with a sense of 'spirits, gods, and other subtle bodies' (Keller and Rubenstein 2017, 5). Crucially, for the purpose of this discussion, it charges 'the materiality of [the] cosmos with ethical intensity, with the spirit of specifically mattering bodies and with the chance of a cosmopolitics of the earth' (Keller and Rubenstein 2017, 3). This is not an overstatement of Carlyle's own position. Whilst the world of men 'was an extinct world, deserted of God', he wrote, it was 'incapable of well-doing' (*John Sterling*, Vol 13, 57). Yet justice would be done, suggests Albert D. Pionke in his own reading of Carlylean Nature, and 'Nature' itself was 'the God-given arbiter of social relations' (Pionke 2001, 9). 'For Carlyle', writes Pionke, 'Nature' is agential, the very point that new materialist thinking has underlined, and its 'agency expresses itself through the execution of God's will in history' (2001, 8). 'It is hostile to human existence', Pionke adds, 'only insofar as humans seek to disobey God' (2001, 8). 'The Universe, I say, is made by Law', Carlyle declared; 'the great Soul of the World is just and not unjust' (Vol 6, 435). Nature was governed by laws; those laws were proof of an absolute, and the unavoidable expression of its will, from which no one could hide.

> In this God's-world, with its wild-whirling eddies and mad foam-oceans, where men and nations perish as if without law, and judgment for an

> unjust thing is sternly delayed, dost thou think that there is therefore no justice? ... I tell thee again, there is nothing else but justice.
>
> (Vol 6, 226)

Here as elsewhere, the stridency of Carlyle's argument betrays its uncertain basis: assuming that the agential reality he calls 'Nature' is God's work, and 'beneficent' (Vol 6, 420), he extrapolates from its workings a form of law that could be called 'justice' (as Geoffrey Tillotson observed) by those who have made 'for themselves a morality which the stars know nothing about' (Tillotson 1978, 80). For many of Carlyle's contemporaries, however, this was a secondary consideration; in an age that was increasingly aware of its own doubts, and eager to find for itself a new and vital form of faith, Carlyle's argument was seized upon because it was rooted in a sense of the divine as manifest within the living world.

The question, nevertheless, is what Carlyle makes of his sense of the divinely natural, and whether his ideas speak in any way to the idea of an emergent justice. As we have seen, Carlyle was alert to the problem of environmental classism, yet sought its solution in the same section of society – an industrial 'aristocracy' – that bore chief responsibility for it. If he was now gesturing towards the intrinsic worth of the nonhuman world, and in so doing 'extending the community of justice' (Coolsaet and Néron 2021, 60) to encompass it, what proposals did he offer that might make good on that gesture? How might society make 'some nook of God's Creation a little fruitfuler, better, more worthy of God'? (Vol 6, 499).

The answer, Carlyle insisted, lay in what he acknowledged to be an 'all-but "impossible" return to Nature, and her veracities' (Vol 6, 439); but it also lay in *work*, and the right to work, '[f]or there is a perennial nobleness, and even sacredness, in Work' (Vol 6, 404). True work, he wrote, '*is* in communication with Nature' (Vol 6, 404); true work lay in (re)establishing the right relationship with 'Nature', a relationship that hinged on the crucial recognition that, in Carlyle's own words, '[t]he Land is *Mother* of us all; nourishes, shelters, gladdens, lovingly enriches us all; in how many ways, from our first wakening to our last sleep on her blessed mother-bosom, does she, as with blessed mother-arms, enfold us all!' (Vol 6, 383). For Carlyle, work lay at the root of an enduring relationship that went beyond the needs of the moment. Indeed, the present was bound to take account of the needs of the future, a view grounded in Carlyle's recognition of what Karen Barad has described as 'the dynamic and contingent materialisation of space, time, and bodies' (2007, 35), and of the fluid entanglements that bind all three together. It is in this sense that Carlyle insisted on the importance of history: it has meaning because it persists in the present and carries forward into the future; it is a continuum, as Carlyle emphasised in *On Heroes, Hero-Worship, and The Heroic in History* (1841 [1897]).

Here, Carlyle made several references to 'Igdrasil', as he spelt Yggdrasil (Nixon 2009, 49), the sacred tree which in Norse mythology acts as a fulcrum for the Nine Worlds.

> Igdrasil, the Ash-tree of Existence, has its roots deep down in the kingdoms of Hela or Death; its trunk reaches up heaven-high, spreads its boughs over the whole Universe: it is the Tree of Existence ... Is not every leaf of it a biography, every fibre there an act or word? Its boughs are Histories of Nations. The rustle of it is the noise of Human Existence, onwards from of old. It grows there, the breath of Human Passion rustling through it ... It is the past, the present, and the future; what was done, what is doing, what will be done ...
>
> (*On Heroes*, Vol 12, 253–4)

As Jude Nixon points out, Carlyle conceived of Igdrasil as 'the articulate voice of diverse creation' (Nixon 2009, 55), and itself '[t]he *Machine* of the Universe' (italics in the original; *On Heroes*, Vol 12, 254), a kind of organic but equally agential alternate to the Protean 'WORLD-MACHINE' (*On Heroes*, Vol 12, 393) of industrial capitalism, whose spatial and temporal reach was now (as Carlyle realised) so extensive. For Carlyle, observes Nixon, 'the manifest difference between the two was that the "world-tree"' Igdrasil connected 'heaven and earth' (Nixon 2009, 50), and was 'constitutive of cosmic harmony and historical continuity' (49).[4] As such, it stood in marked opposition to the temper (and tempo) of the age, that nomadic, 'swift-rolling, self-abrading Time' (Vol 6, 480). 'Permanence, persistence is the first condition of all fruitfulness in the ways of men', wrote Carlyle; civilisation itself depended on 'long-dated hundred-fold possessions, not to be valued in the money-market', and on 'memories and hopes, even for this Earth, that reach over thousands of years' (Vol 6, 480). True work lay in recognising that dependency, and cultivating it; it lay in an accommodation between humankind and the earth itself, a long-term accommodation which would restore to both their dignity and their divinity.

> Here let us rest, and lay-out seedfields; here let us learn to dwell. Here, even here, the orchards that we plant will yield us fruit; the acorns will be wood and pleasant umbrage, if we wait. How much grows everywhere, if we do but wait! Through the swamps we will shape causeways, force purifying drains; we will learn to thread the rocky inaccessibilities; and beaten tracks, worn smooth by mere travelling of human feet, will form themselves.
>
> (Vol 6, 483)

The working relationship that Carlyle described reflects this sense of the 'long-term imbrication' of people and place – itself a critical definition of what it is to dwell in a mindful, environmentally sustainable way (Garrard 2012, 117) – and the recognition of intergenerational equity that it entails; it recognises and reflects the complex ways in which the past informs the present, and both flow into the future. The land, wrote Carlyle, 'is not the property of any generation ... but that of all the past generations that have worked on it, and of all the future ones that shall work on it' (Vol 6, 384). Thus, Carlyle's view of the land reflected his strong sense of its value independent of the cash 'nexus' and as only in part a product of the labour invested in it. Just as the work done on it 'is of a religious nature' (Carlyle Vol 6, 407), so the land is itself sacred, and the right relationship to it is not one of ownership but one of stewardship, as Carlyle made clear in *Past and Present*.

> Men talk of 'selling' Land. Land, it is true ... in such a trading world, has to be presented in the market for what it will bring, and as we say be 'sold': but the notion of 'selling', ... is a ridiculous impossibility! We buy what is saleable of it; nothing more was ever buyable. Who can or could sell it to us? Properly speaking, the Land belongs to these two: To the Almighty God; and to all His Children of Men that have ever worked well on it, or that shall ever work well on it ... Ah yes, soil, with or without ploughing, is the gift of God. The soil of all countries belongs evermore, in a very considerable degree, to the Almighty Maker! The last stroke of labour bestowed on it is not the making of its value, but only the increasing thereof.
>
> (Vol 6, 384–5)

Through this emphasis on the intrinsic worth of the land, and on the importance of a working relationship with it, Carlyle opened up a vision of what David Fairer (2011) has described (drawing on a long tradition of writing about farming) as an eco-Georgic: a non-hierarchical relationship to the environment, embodying a 'mutual respect between man and nature' (Fairer 2011, 202), alert to the dynamism and the 'stubborn materiality' (206) of a constantly changing world, and the ceaseless labour demanded by its cultivation. The land, argued Carlyle, is held in trust for the next generation by those who labour on it; it is to the generations that follow – and, ultimately, to the God who first gifted it to humankind – that the land ultimately belongs; and necessarily, this relationship must be an attentive, mindful one, cognisant that the land has a value far beyond the price that might be put upon it.

Problematically, however, Carlyle's argument does not end here, in a conclusion that might align his views with an ecological critique of industrial capitalism and its wider impacts, or of the social, environmental, and species injustices that are its result. Tensions and contradictions remain. As an antidote to modern ills, Carlyle suggested a return to the land, where citizen farmers might then till the soil in a Georgic relationship of long-term responsibility – an influential proposition, which was taken up by the Chartist movement in the 1840s (Gould 1988, 6–9) and later in the century by a range of 'back-to-the-land movements' (59–75). But whilst Carlyle implied that the land should belong to those who labour on it, which would entail a truly radical transformation of land ownership, he nowhere challenged the existing status quo. Nor did he acknowledge that 'the Land' is part of a larger, living system, with which the flourishing of other, nonhuman species is equally entangled; Carlyle's 'Nature' was, more often than not, an abstraction from which natural processes are absent, and the nonhuman is incidental, without any claim of its own. Moreover, and whilst 'the Land' might remain 'evermore' that of the Almighty – a gift that is only ever (as it were) on loan – Carlyle was (as we have seen) enthralled, perhaps even dazzled by the power of the machine to supplement or even supplant human labour, and in so doing subjugate the very 'Nature' he elevated to the level of the divine.[5] Furthermore, and notwithstanding his trenchant critique of the insidious effects of a Machine Age, he regarded this extension of human power over the nonhuman world as itself the expression of God's will, and perhaps the only positive consequence of the epochal transformation of Victorian Britain. 'With all its miserable shortcomings, with its wars, controversies, with its trades-unions, famine-insurrections', Carlyle wrote, 'it is her Practical Material Work alone that England has to show for herself!' (Vol 6, 378).

> The grim inarticulate veracity of the English people, unable to speak its meaning in words, has turned itself silently on things; and the dark powers of Material Nature have answered, 'Yes, this at least is true, this is not false!' So answers Nature … Mountains, old as the Creation, I have permitted to be bored through; bituminous fuel-stores, the wreck of forests that were green a million years ago, – I have opened them from my secret rock-chambers, and they are yours, ye English … So answers Nature. The Practical Labour of England is *not* a chimerical Triviality: it is a Fact, acknowledged by all the Worlds; which no man and no demon will contradict. It is, very audibly, though very inarticulately as yet, the one God's Voice we have heard in these two atheistic centuries.
>
> (Italics in the original; Vol 6, 378)

Carlyle was not alone in finding this transformation dazzling, then or since. Societies have everywhere emulated it, as the historian L. C. B. Seaman remarked (1973 [1993], 26). Nor is it surprising that Carlyle failed to foresee the kinds of problems that this wholesale exploitation has since generated; it is only in recent years that the developed world has, for example, begun to acknowledge the impact of its opening up of those 'bituminous fuel-stores' (i.e. coal, still today the principal energy source on the planet; Munawer 2018, 87). But it is surprising that Carlyle should see no limits to this restless and continuous process of expansion, or ask what might come of transgressing those limits. Was 'Nature' not, after all, a gift from God, within which God's presence might be felt? The irony of Carlyle's position is that, whilst he recognised a vital life-force at work in the vast immensity of space and time, aligned that force with his own sense of the divine, and identified in its laws the very basis of a sense of justice, he did not ask if humankind's ever more relentless exploitation of that world might itself constitute a kind of trespass or offence, or itself an injustice; whether, if there were limits to that process of ever more efficient exploitation, it might one day rebound on humankind. *That* question would find its highly imaginative answer in the speculative fiction of the late-Victorian period, when writers began to create extravagant fantasies in which 'Nature' did indeed recoil from human interference – and revenge itself on an errant species.[6]

*

By the time Carlyle posed the Condition-of-England Question, the problem of environmental classism – and the disquiet to which it gave rise – was already a growing concern, even amongst those within society who had hitherto preferred to ignore its existence. Carlyle was no less troubled by this convergence of poverty and pollution, and equally convinced that it was unjust. In Carlyle's work, however, these are only ever touchstones for a much more probing analysis of what he diagnosed as 'a mighty change in our whole manner of existence' (Vol 14, 468). Unfettered forms of capitalism and industrialisation, the subtler and more pervasive effects of industrialism, and the deadening rationalism of Enlightenment thinking had together created a 'Human Chaos' (Vol 6, 473). As Carlyle stridently insisted, political and juridical processes were themselves at fault. The law was a labyrinth, a point with which Charles Dickens would later engage, most memorably in the novel *Bleak House* (1852–3; see Chapter 4); parliament was simply self-serving. But whilst both aristocracy and radicals had failed to represent working-class concerns or enact the kind of legislation that might ameliorate working-class suffering, Carlyle nevertheless presented the working class as themselves incapable of self-representation. In his view, they were 'dumb', inarticulate creatures, and they would remain inferior, and ever in need of guidance and government by their 'Real-Superiors' (Vol 6, 427). Thus, Carlyle identified the link between environmental inequality and a lack of political involvement and recognition, only to gainsay any possibility of participation through his own *mis*recognition of this marginalised and impoverished class. Democracy was not, in his opinion, any kind of solution. Unsurprisingly, his own answers to the problem of working-class marginalisation and exploitation – such as emigration and better education (Vol 6, 466, 470) – were reformist rather than revolutionary; indeed, it was fear of revolution that, after 1848, shaped the 'negative and authoritarian worldview' (Heyrendt-Sherman 2021, 168) of his later work.

Society itself chose a reformist path, diverting rather than directly addressing the animus to which Carlyle's great works of social commentary gave voice.[7] But perhaps the greater part of his message – and the part of it that resonated most deeply with a generation of Victorians whose own faith had been undermined by the rationalising discourse that Carlyle critiqued – lay in his finding of a kind of faith in the living spirit of 'Nature'. If the prophetic tone and declamatory urgency of this message owed a great deal to Calvinism, the message itself owed just as much to the literary phenomena of German Romanticism. Indeed, Carlyle's enormous influence on his contemporaries underlines the point that, as he himself observed, literature '*is* yet "the Thought of Thinking Souls"' (italics in the original; Vol 6, 442).

In Carlyle's work, a divine presence is the bedrock of all being, and of human being; the universe and its laws (and hence justice itself) are themselves an expression of that presence. It is in this sense of the world as something more than dead matter, awaiting human intervention for its value to be realised, that Carlyle came closest to acknowledging the need for a wider, emergent justice, beyond the species-limit of human concerns. This is also the point at which Carlyle's critique of environmental injustice stumbled over its own inherent contradictions. The earth was a gift from God, given to humankind for it to work, but also master and subjugate; to make fruitful, by any and all means; and so 'the grand Industrial task of conquering [this] Terraqueous Planet for the use of man' (Vol 16, 258) was, like cotton-spinning itself, sacred, 'noble' (Vol 6, 363), and the sound of Manchester's cotton mills 'sublime as a Niagara' (Vol 16, 264).

In consequence, Carlyle did not equate the environmental inequalities that overshadowed the working class with the growing degradation and exploitation of the natural world. Rather, he valorised the Promethean triumph of humankind over nature as itself an extension of the imperative to work, and assumed that there was no limit to the earth's capacity to absorb the demands of an industrialised society: 'the world is wide enough yet for another Six Thousand Years', he wrote, in reference to the Biblical calculations of the world's age (Vol 6, 470–1). In his mind, the question was only whether Victorian Britain could accomplish 'the grand Constitutional task of sharing, in some pacific endurable manner, the fruit of said conquest' (Vol 16, 258), and in so doing, resolving the iniquities and tensions that had given rise to Chartism.

Hence it might be argued that, as George Meredith later observed, Carlyle was a 'heaver of rocks and not a shaper' (1912, 333), an iconoclast whose critique of the impact of industrial modernity was more persuasive than the solutions he proposed to them, and who drew back from the logical implications of his analysis of 'the condition and disposition of the Working Classes' (Carlyle Vol 16, 204): that revolution was not only possible, but necessary. The next chapter turns to the work of Friedrich Engels, 'one of the most important socialist and ecological thinkers in human history' (Clark and Foster 2006, 376). Engels drew heavily on Carlyle's analysis of the Condition of England; unlike Carlyle, however, he believed that revolution was both a historical necessity and a historical inevitability, which would in turn lead to the creation of a more equitable and free society. The roots of that belief lay in his own, first-hand encounters with the 'sordid reality' (Piketty 2014 [2017], 9) created by industrial modernity. As I discuss, those encounters formed the basis of *The Condition of the Working Class in England* (1845), a defining account of environmental classism in early Victorian Britain (Bell 2020, 2).

*

Notes

1 The copy text for Carlyle's writings is *The Works of Thomas Carlyle* (New York, NY: Peter Fenelon Collier, 1897) in 16 volumes, available at archive.org. In-text citations refer to the relevant volume number, with the short title of the individual work, e.g. 'Latter-Day Pamplets', in those cases where works occupy the same volume.

2 The health and safety risks of exposure to copperas, or iron (II) sulfate heptahydrate, are set out in safety data sheets, now widely available online (see, for example, Fisher Scientific 2021). Byssinosis was sufficiently widespread and well-known for Carlyle to be aware of it, yet it was not officially mentioned until 1861 (Gill 1947, 48). According to C. I. C. Gill, writing over a century later, no research was undertaken into the disease 'until recently', perhaps because 'the cotton trade has been the Cinderella of occupational hazards' (1947, 48).

3 Almost from the outset, Carlyle's critics referred to 'Carlylese' as a descriptor of his highly mannered prose; over time, and as that prose grew increasingly violent and excessive – and Carlyle's views steadily more authoritarian, reactionary, and intolerant – it also became a highly critical one; for a detailed discussion see Reitz 1997.

4 As Nixon points out, Carlyle wrote an essay on the subject of Igdrasil which was posthumously published 'in the February 1890 issue of *Igdrasil: Journal of the Ruskin Reading Guild*, a title that reflects the Yggdrasil mania of the nineteenth century advanced largely by Carlyle and Ruskin' (Nixon 2009, 53).

5 For a more detailed discussion of Carlyle's incongruent positioning of Nature as itself the expression, even embodiment of God, see Albert D. Pionke (2001), who notes that 'human efforts to manipulate the environment would represent [by definition] a clearly unacceptable act' (12); thus, the positioning of the human as the legitimate exploiter of Nature is a 'dramatic reversal' of Carlyle's earlier arguments, a reversal that makes possible suggestions for reform, such as emigration, which would avoid the Malthusian trap of starvation by allowing an exponentially growing population free access to hitherto untouched lands (Pionke 2001, 16).

6 M. P. Shiel's remarkable 'last man' novel, *The Purple Cloud* (1901), is an instance. Shiel's novel was written at the very end of Victoria's reign, and published just after her death, an event which to some, at least, might well have seemed world ending. In Shiel's novel, apocalypse is triggered by human trespasses, perhaps against Yggdrasil itself. 'So answers Nature', in Carlyle's own words (Vol 6, 378). Humankind can hardly claim it has not been warned; the novel features a wild-haired Scottish preacher whose exhortations sound very much like Carlyle in full voice.

7 These reformist measures took various forms, including better regulation of factory employment, sanitary interventions, constraints on child labour, and civic improvements to allow for rational recreation; for example, Preston's Moor Park (1844) 'was one of the first of a wave of municipal amenities' (Hilton 2006, 631),

notable, of course, because it reflected the growing recognition that an environment from which so many manifestations of nonhuman and natural life had been erased or excluded was in and of itself harmful to human well-being. As Dorothy and Alan Shelston point out (1990, 67), these reforms were of a piece with those that Carlyle had himself suggested.

Works cited

Alaimo, Stacey. 2010. *Bodily Natures: Science, Environment, and the Material Self*. Bloomington, IN: Indiana University Press.

Barad, Karen. 2007. *Meeting the Universe Halfway: Quantum Physics and the Entanglement of Matter and Meaning*. London: Duke University Press.

Bell, Karen. 2020. *Working-Class Environmentalism: An Agenda for a Just and Fair Transition to Sustainability*. Cham, Switzerland: Palgrave Macmillan.

Buckley, Jerome Hamilton. 1952 [1966]. *The Victorian Temper: A Study in Literary Culture*. London: Frank Cass.

Carlyle, Thomas. 1843 [1897]. *Past and Present. The Works of* Thomas *Carlyle, Volume Six*, 417–500. New York, NY: Peter Fenelon Collier. Available at archive .org, https://archive.org/details/worksofthomascar0006carl/page/n5/mode/2up ?view=theater. Accessed 15 November 2022.

Carlyle, Thomas. 1833–4 [1897]. *Sartor Resartus: The Life and Opinions of Herr Teufelsdröckh. The Works of* Thomas *Carlyle, Volume Twelve*, 1–225. New York, NY: Peter Fenelon Collier. Available at archive.org, https://archive.org/details/carlyle completew12carl/page/n7/mode/2up?view=theater. Accessed 15 November 2022.

Carlyle, Thomas. 1841 [1897]. *On Heroes, Hero-Worship, and The Heroic in History. The Works of* Thomas *Carlyle, Volume Twelve*, 233–461. New York, NY: Peter Fenelon Collier. Available at archive.org, https://archive.org/details/wor ksofthomascar0012carl/page/n5/mode/2up. Accessed 17 November 2022. These lectures were delivered in 1840, and first published in 1841.

Carlyle, Thomas. 1850 [1897]. 'Latter-Day Pamphlets'. *The Works of* Thomas *Carlyle, Volume Thirteen*, 261–455. New York, NY: Peter Fenelon Collier. Available at archive.org, https://archive.org/details/worksthomascarl38carlgoog/ page/n10/mode/2up?view=theater. Accessed 17 November 2022.

Carlyle, Thomas. 1851 [1897]. *The Life of John Sterling. The Works of* Thomas *Carlyle, Volume Thirteen*, 1–258. New York, NY: Peter Fenelon Collier. Available at archive.org, https://archive.org/details/worksofthomascar0013carl/page/n5/ mode/2up?view=theater. Accessed 16 November 2022.

Carlyle, Thomas. 1829 [1897]. 'Signs of the Times'. *The Works of* Thomas *Carlyle, Volume Fourteen*, 462–87. New York, NY: Peter Fenelon Collier. Available at archive.org, https://archive.org/details/worksthomascarl07carlgoog/page/n12/ mode/2up?view=theater. Accessed 17 November 2022.

Carlyle, Thomas. 1831 [1897]. 'Characteristics'. *The Works of* Thomas *Carlyle, Volume Fifteen*, 208–47. New York, NY: Peter Fenelon Collier. Available at archive.org, https://archive.org/details/carlylecompletew15carl/page/n7/mode/2up ?view=theater. Accessed 17 November 2022.

Carlyle, Thomas. 1839 [1897]. *Chartism. The Works of* Thomas *Carlyle, Volume Sixteen*, 204–85. New York, NY: Peter Fenelon Collier. Available at archive.or g, https://archive.org/details/worksofthomascar0016carl/page/n7/mode/2up?view =theater. Accessed 16 November 2022.

Clark, Brett, and John Bellamy Foster. 2006. 'The Environmental Conditions of the Working Class: An Introduction to Selections from Frederick Engels's *The Condition of the Working Class in England in 1844*'. *Organization & Environment* 19 (3): 375–88. Available at jstor.org, www.jstor.org/stable/26162420. Accessed 22 November 2022.

Cook, E. T., and Alexander Wedderburn (eds). 1903–12. *The Library Edition of John Ruskin's Works.* 39 vols. London: George Allen. Available at www.lancaster.ac.uk/the-ruskin/the-complete-works-of-ruskin/. Accessed 5 January 2023.

Coolsaet, Brendan, and Pierre-Yves Néron. 2021. 'Recognition and environmental justice'. *Environmental Justice: Key Issues*, edited by Brendan Coolsaet, 52–63. London: Routledge.

Daunton, Martin. 2000. 'Society and economic life'. *The Nineteenth Century*, edited by Colin Matthew, 41–82. Oxford: Oxford University Press.

Eliot, George. 1971 [2002]. 'An unsigned review'. Thomas Carlyle: *The Critical Heritage*, edited by Jules Paul Seigel, 409–11. London: Routledge.

Faber, Daniel. 2018. 'The political economy of environmental injustice'. *The Routledge Handbook of Environmental Justice*, edited by Ryan Holifield, Jayajit Chakraborty, and Gordon Walker, 61–73. Abingdon: Routledge.

Fairer, David. 2011. ''Where Fuming Trees Refresh the Thirsty Air:' The World of Eco-Georgic'. *Studies in Eighteenth-Century Culture* 40: 201–18. doi: 10.1353/sec.2011.0006

Fisher Scientific. 2021. *Safety Data Sheet: Iron(II) sulfate heptahydrate*, 26 December. Available at www.fishersci.com/store/msds?partNumber=AC423730050&productDescription=IRON%28II%29+SULFATE+HEPTAHYDR+5G&vendorId=VN00032119&countryCode=US&language=en. Accessed 9 February 2022.

Fraser, Hilary, and Daniel Brown. 1997. *English Prose of the Nineteenth Century.* London: Longman.

Froude, James Anthony. 1882. *Thomas Carlyle: A History of the First Forty Years of his Life*, Vol. II. New York, NY: Charles Scribner's Sons. Available at archive.org, https://archive.org/details/05837939.18547.emory.edu/page/n3/mode/2up?q. Accessed 7 December 2020.

Garrard, Greg. 2012. *Ecocriticism*. 2nd edn. Abingdon: Routledge.

Gill, C. I. C. 1947. 'Byssinosis in the Cotton Trade'. *British Journal of Industrial Medicine* 4 (1): 48–55. Available at jstor.org, www.jstor.org/stable/27720626. Accessed 9 February 2022.

Gould, Peter C. 1988. *Early Green Politics: Back to Nature, Back to the Land, and Socialism in Britain, 1880–1900*. Sussex: Harvester Press.

Heyrendt-Sherman, Catherine. 2021. 'From Social Reformer to Political Conservative: Thomas Carlyle and the Practicalities of French and Victorian Politics'. *Carlyle Studies Annual* 34: 167–88. Available at jstor.org, www.jstor.org/stable/27169822. Accessed 18 November 2022.

Hilton, Boyd. 2006. *A Mad, Bad, and Dangerous People? England 1783–1846.* Oxford: Clarendon Press.

John, Brian. 1974. *Supreme Fictions: Studies in the Work of William Blake*, Thomas Carlyle, *W. B. Yeats, and D. H. Lawrence.* Montreal: McGill-Queen's University Press.

Keller, Catherine, and Mary-Jane Rubenstein. 2017. 'Introduction: Tangled Matters'. *Entangled Worlds: Religion, Science, and New Materialisms*, edited by Catherine

Keller and Mary-Jane Rubenstein, 1–18. New York, NY: Fordham University Press.

Keller, Catherine. 2017. 'Tingles of Matter, Tangles of Theology'. *Entangled Worlds: Religion, Science, and New Materialisms*, edited by Catherine Keller and Mary-Jane Rubenstein, 111–35. New York, NY: Fordham University Press.

Marx, Karl, and Friedrich Engels. 2002. *The Communist Manifesto*, introduced by Gareth Stedman Jones. London: Penguin.

Marx, Karl. 1975 [1992]. 'Economic and Philosophical Manuscripts (1844)'. *Early Writings*, translated by Rodney Livingstone and Gregor Benton, 279–400. London: Penguin.

Maurice, Frederick (ed.). 1885. *The Life of Frederic Denison Maurice, Chiefly Told in his Own Letters, Edited by his Son, vol I*. London: Macmillan. Available at archive.org, https://archive.org/stream/lifeoffrederickd01mauriala/lifeoffrederickd01mauriala_djvu.txt. Accessed 6 December 2020.

Meredith, George. 1912. *Letters of George Meredith: Volume II*. London: Constable. Available at archive.org, https://archive.org/details/in.ernet.dli.2015.173828/page/n3/mode/2up?q=heaver. Accessed 4 December 2020.

Morton, Timothy. 2010. *The Ecological Thought*. London: Harvard University Press.

Morton, Timothy. 2013. *Hyperobjects: Philosophy and Ecology after the End of the World*. London: University of Minnesota Press.

Morton, Timothy. 2016. *Dark Ecology: For a Logic of Future Coexistence*. Chichester: Columbia University Press.

Munawer, Muhammad Ehsan. 2018. 'Human health and environmental impacts of coal combustion and post-combustion wastes'. *Journal of Sustainable Mining* 17 (2): 87–96. doi: 10.46873/2300-3960.1125

Nixon, Jude V. 2009. 'Thomas Carlyle's Igdrasil'. *Carlyle Studies Annual* 25: 49–58. Available at jstor.org, www.jstor.org/stable/26593165. Accessed 8 December 2020.

Palma, Vittoria Di. 2014. *Wasteland: A History*. New Haven, CO: Yale University Press.

Parham, John. 2010. *Green Man Hopkins: Poetry and the Victorian Ecological Imagination*. Amsterdam: Rodopi.

Pellow, David Naguib. 2018. *What is Critical Environmental Justice?* Cambridge: Polity.

Piketty, Thomas. 2014 [2017]. *Capital in the Twenty-First Century*, translated by Arthur Goldhammer. Cambridge, MA: Belknap Press. First published in 2013.

Pionke, Albert D. 2001. 'Carlylean Nature'. *Carlyle Studies Annual* 20: 7–26. Available at jstor.org, www.jstor.org/stable/44945800. Accessed 7 December 2020.

Plumwood, Val. 1993 [1997]. *Feminism and the Mastery of Nature*. London: Routledge.

Proudhon, Pierre-Joseph. 1847 [2007]. *The Philosophy of Misery*. New York, NY: Cosimo.

Reitz, Caroline. 1997. 'Beneath Barking and Froth: Carlyle and the Othering of Violence'. *Carlyle Studies Annual* 17: 7–21. Available at jstor.org, www.jstor.org/stable/44945650. Accessed 21 November 2022.

Seaman, L. C. B. 1973 [1993]. *Victorian England: Aspects of English and Imperial History, 1837–1901*. London: Routledge.

Shelston, Dorothy, and Alan Shelston. 1990. *The Industrial City 1820–1870*. Houndmills: Basingstoke.

Swift, Roger. 2001. 'Thomas Carlyle, 'Chartism', and the Irish in Early Victorian England'. *Victorian Literature and Culture* 29 (1): 67–83. Available at jstor.org, www.jstor.org/stable/25058540. Accessed 7 December 2020.

Tillotson, Geoffrey. 1978. *A View of Victorian Literature*. Oxford: Clarendon.

Willey, Basil. 1949 [1973]. *Nineteenth-Century Studies: Coleridge to Matthew Arnold*. Harmondsworth: Pelican.

2 Friedrich Engels, environmental classism, and 'social murder'

In Britain, industrialisation was cotton-led, and 'Cottonopolis' was Manchester, the symbol and 'shock-city' of this new industrial age (Hilton 2006, 578). Inevitably, Manchester became the object of often anxious inquiry, as a succession of visitors, reporters, writers, and investigators inquired into its changing circumstances and sought to understand its wider implications. To Benjamin Disraeli, writing in 1844, it was 'the most wonderful city in modern times' (1844 [1982], 181); to Alexis de Tocqueville, in 1835, it was also one in which 'civilised man is turned back almost into a savage' (1835 [1958], 108). In turn, Manchester became the focus of one of the most remarkable and ultimately influential accounts of environmental classism in Victorian Britain: Friedrich Engels's *The Condition of the Working Class in England*, written between 1842 and 1844.

In *The Condition of the Working Class*, Engels provided a graphic account of the environmental inequities generated by industrial modernity and their transformation into a structural form of environmental classism. Grounded in first-hand observation, Engels's 'scathing, ruthlessly documented portrayal' (Clark and Foster 2006, 376) still bears scrutiny as a historical record of industrial modernity's impact (Cannadine 2017, 207). Notwithstanding its reputation as a factual record – and its later standing as a foundational, empirical element in Marxist thinking (Sherwood 1985, 837) – *The Condition of the Working Class* is also a consciously literary artefact. As I discuss in more detail below, Engels's pioneering account of environmental classism is all the more compelling and moving because he weaves into it elements of two 'deeply entrenched' literary tropes (Garrard 2012, 37), the pastoral and the Gothic. The first of these tropes forms part of Engels's opening portrayal of unalienated life and labour before the advent of industrial modernity; the second is woven into his depiction of the blighted industrial underworld with which it is then juxtaposed. From a new materialist perspective, this contrast is all the more important because it gestures towards an expanded awareness of the connections that make for a truly flourishing life; it gestures, in other words, towards what I have tentatively called an emergent justice, which takes account of nonhuman as well human flourishing. The world that Engels describes in his description of a green, leafy, and 'Merry Old

DOI: 10.4324/9780367821609-3

England' (Engels 1993 [2009], 34)[1] is, in the terms that Karl Marx was then developing in his 'Economic and Philosophical Manuscripts' (1844), an unalienated one, an embodied existence 'predicated on love for objects or beings outside of ourselves, which necessarily requires their continued existence' (Foster and Burkett 2016 [2017], 55). Here, nature and culture are entangled, and unthinkingly so; this unspoken (inter)relationship has itself come to seem natural, inevitable. With the advent of industrial modernity, however, that relationship is strained and then severed, as the working classes are alienated from nature and from their own natures. Matter increasingly appears as agential, generative; as itself independent, incomprehensible, and ultimately, threatening. Ulrich Beck's 'risk consciousness' (1992 [2012], 72) is also, therefore, consciousness of alienation, and of an alienating world.

Significantly, the Victorian's growing presentiment of a world filling with misbehaving, alien entities anticipates modern forms of speculative realism such as object-oriented ontology (Gratton 2014 [2015], 85–133), from which Timothy Morton derived his influential theory of hyperobjects (Morton 2013, 2014), entities that (like industrial modernity itself) are at once present yet permanently withdrawn, and which exist at or beyond the (spatial and temporal) threshold of human comprehension, even imagination. New materialist thinking suggests a further interpretation. As Karen Barad outlines in her theory of an agential reality (2007, 32–5), entanglement is inescapable because it is constitutive: it is only through entanglement that entities emerge as such. '[D]istinct agencies do not precede', she argues, 'but rather emerge through, their *intra-action*' (italics added), a neologism she uses to denote this mutual entanglement (Barad 2007, 33). Thus, intra-action precedes separability, but separability presupposes (further) intra-activity, in a generative process that is always continuing, always iterative. In this sense, there is only ever the 'naturalcultural', as itself a material-discursive formation whose meaning is mutable and elusive (Barad 2007, 32).

With the advent of industrial modernity, however, this (inter)relationship changes, as the cultural increasingly dominates the natural, creating the perception that the nonhuman or natural world is in jeopardy, in question. 'Nature' emerges as a category in itself, marking its apparent loss or withdrawal, as society is increasingly absorbed into the built environments generated by industrial modernity, and long-standing and sustainable relationships are supplanted by exploitative ones, in which 'Nature's' conquest is itself a goal, and source of pride. Now, the naturalcultural is characterised by the relative subjugation or even absence of the nonhuman; indeed, the very term 'naturalcultural' implies a balance that that 'does not do justice to ecological trauma', argues Jesse Oak Taylor (2016, 5). New language is needed to register these effects; as I discuss further below, Taylor's own neologism, the 'abnatural' (Taylor 2016, 5), provides a useful tool for understanding Engels's wider purpose, in describing a world in which the impact of industrial modernity is not confined to the short-term, the localised, or even the human, but insidiously works its way outwards to encompass and undermine the very basis of a flourishing life.

*

To the young Friedrich Engels (1820–1895), newly arrived in Manchester, British society was in the midst of an epochal change, and 'Cottonopolis' was simply the most advanced stage of a process that was already transforming society as a whole. Recording his impressions and emerging ideas, Engels did not therefore confine himself to descriptions of specific, localised instances of environmental injustice, although his narrative everywhere seeks to particularise the changes he is describing and the problems those changes are creating. Like Carlyle before him, he is concerned with the more profound problem of what David Pellow describes as 'environmental inequality' (2000, 582), and the way in which the 'deeper workings of power in society' (597) result in both 'the unequal distribution of power and resources' and of environmental harms or 'burdens' (582). As Pellow adds, injustice on this scale goes beyond localised 'sacrifice zones' to encompass 'entire *populations*' (italics in the original; Pellow 2018, 17): in Victorian Britain, it was the working (or 'labouring') classes which were primarily (but not exclusively) subjected to this process of 'ideological and institutional othering' (2018, 17), and it is in this sense that we can talk not simply of environmental inequality, but of environmental classism, the issue with which Engels is himself fundamentally concerned.[2]

For Engels, the creation of this 'expendable' (Pellow 2018, 7) grouping within society – a creation that constituted, as he recognised, a revolution in social relations – was not just a by-product of technological (and specifically industrial) innovation, but the structural concomitant of an intense and historically unprecedented process of capital accumulation (Piketty 2014 [2017], 1). Politics played no part in it; or rather, and as Eric Hobsbawm would later observe, politics itself was 'already geared to profit' (1975 [1995], 31). 'All the conditions of life are measured by money', wrote Engels (282); 'as Carlyle says ... "Cash Payment is the only nexus between man and man"' (282):

> Hence free competition in every respect, hence the regime of *laissez-faire, laissez-aller* in government ... Free competition will suffer no limitation, no State supervision; the whole State is but a burden to it. It would reach its highest perfection in a wholly ungoverned anarchic society, where each might exploit the other to his heart's content.
>
> (Italics in the original; 282–3)

As Engels also realised, the triumph of industrial modernity required not only the absence of state interference, but the creation of a subordinate class, entirely dependent for its survival on the sale of its own labour. This structural transformation enabled one class to dominate and instrumentalise the other, turning people into fractions of themselves: the bourgeoisie, Engels argued, saw in the working classes 'not human beings, but hands, as [the mill or factory owner] constantly calls them to their faces' (282). Like the natural world that industrial modernity was increasingly able to exploit, the broad mass of society had simply become a resource, valuable only to the extent that it was useful. It was, as Thomas Piketty observes, a system that knew 'neither limits nor morality' (2014 [2017], 7).

Against this backdrop, Engels focused his attention on the fate of the industrial proletariat, and the centrality of city life to the working classes. 'Since commerce and manufacture attain their most complete development in [the] great towns', he wrote, 'their influence upon the proletariat is also most clearly observable here' (1993 [2009], 34). Manchester was his exemplum; here, environmental inequality was embodied in the growing 'psycho-social' divide between the working poor and their affluent masters (Krishnamurthy 2000, 427). As Engels pointed out, the rich lived in newly built suburbs, where they were able to enjoy 'wholesome country air, in fine, comfortable homes' (58), out of reach of slum squalor and its disturbing implications;

> [a]nd the finest part of [this systematic segregation is] that the members of [the] money aristocracy can take the shortest road through the middle of all the labouring districts to their places of business, without ever seeing that they are in the midst of the grimy misery that lurks to the right and the left.
>
> (58)

Meanwhile, added Engels, 'the working-people's quarters are sharply separated from the sections of the city reserved for the middle class' (57). Here, the poor eked out their lives amidst 'misery and grime' (58). Houses were 'dirty, old, and tumbledown'; newer buildings were crammed into the remaining space, creating a 'tangle into which they are crowded literally one upon the other' (60). Amidst the narrow alleys, covered passages, and courts of these districts, 'filth and disgusting grime' were inescapable (61): '[i]n one of these courts', wrote Engels,

> there stands directly at the entrance, at the end of the covered passage, a privy without a door, so dirty that the inhabitants can pass into and out of the court only by passing through foul pools of stagnant urine and excrement.
>
> (61)

Mills were interspersed with dwellings (60); tanneries filled the neighbourhoods with 'the stench of animal putrefaction' (60). The 'drains and refuse' of 'tanneries, bonemills, and gasworks', Engels observed, 'all find their way into the [adjacent river] Irk, which receives further the contents of all the neighbouring sewers and privies' (62). 'Everywhere', Engels added, 'heaps of debris, refuse, and offal; standing pools for gutters, and a stench which alone would make it impossible for a human being in any degree civilised to live in such a district' (63). 'Such is the Old Town of Manchester', Engels concluded, 'containing at least twenty to thirty-thousand inhabitants', a place in which everything that 'arouses horror and indignation is of recent origin, belongs to the *industrial epoch*' (italics in the original; 65).[3]

As Engels describes this and other slum districts in the city, it becomes clear that the conditions he is describing frequently resemble, anticipate, and sometimes far exceed the environmental discrimination and environmental inequalities with which environmental justice activists and scholars have engaged in the last few decades. Segregated in slums, the working classes lived side-by-side with polluting industries in dilapidated, unplanned, and overcrowded accommodation. Lacking almost any form of environmental regulation or protection, they were exposed to environmental health hazards ranging from human waste to industrial pollutants, and to the ailments and diseases to which these hazards give rise. But as Engels's account underlines, those hazards were not confined to the conditions of life in the slums. Rather, working-class communities found themselves at the intersection of what environmental justice scholar Daniel Faber describes as a '"quadruple exposure effect" to environmental health hazards' (2018, 62): not only were their neighbourhoods more polluted, but their places of work were more dangerous, their 'household, commercial foods and a variety of consumer products' were more likely to contain toxic materials, and in the turbulent climate created by capitalism, they were likely to 'face greater dislocation, health problems, and loss of livelihood' (2018, 63).

These concerns also form part of Engels's narrative, itself written during one of the most intense economic depressions of the age (Sherwood 1985, 837). For example, Engels described at length the dangers of work life in the new factories. Like Carlyle, he noted the impact of breathing the 'fibrous dust' generated in cotton and flax-spinning mills (172), a widespread condition that would later be called byssinosis (Gill 1947, 48). In addition, Engels described the 'multitudes of accidents' (173) caused by working with fast-moving and unguarded machinery (174). Sometimes, the injuries were minor, but the consequences were not, since lockjaw (or tetanus) invariably resulted, with often fatal consequences (173): no cure or vaccine was available, a point which relates directly to the conditions created by a risk society, in which industrial modernity creates new, sometimes invisible, and often poorly-understood threats; some of the more insidious illnesses associated with cotton-manufacture, such as mule spinner's cancer, were not identified for decades (see Waldron 1983). Accidents, injuries, diseases, and 'malformations' (173) were, concluded Engels in a characteristically forceful denunciation, the inevitable result of these working conditions:

> A pretty list of diseases engendered purely by the hateful money-greed of the manufacturers! Women made unfit for child-bearing, children

> deformed, men enfeebled, limbs crushed, whole generations wrecked, afflicted with disease and infirmity, purely to fill the purses of the bourgeoisie.
>
> (175)

There were, as Engels carefully noted, other concerns. It was difficult to obtain wholesome food, when much of what was available was tainted or doctored, and it was often difficult to find adequate clothing (Clark and Foster 2006, 385). Since wages were paid at a subsistence level (Clark and Foster 2006, 382), workers found it difficult if not impossible to save, the supposed solution to many of their complaints. Without the opportunity to build up financial reserves, they were particularly exposed to the impacts of illness, injury, or unemployment. For families who were already on the breadline, any one of these conditions could result in near-starvation conditions; because of the filth and squalor of slum life, these malnourished communities were also disproportionately vulnerable to the diseases whose spread was itself engendered by the conditions in which lived. Thus, the working classes of Engels's account were caught in a trap at once lethal and toxic, created by multiple, converging forms of environmental discrimination. This was, Engels argued, a form of 'social murder', everywhere recognised, but nowhere acknowledged;

> murder against which none can defend himself, which does not seem what it is, because no man sees the murderer, because the death of the victim seems a natural one, since the offence is more one of omission than of commission.
>
> (106–7)

As Engels's account of life in Manchester's slums also reveals, environmental racism was a factor in this discriminatory construct. Several working-class districts in Manchester were populated by Irish immigrants, to whom Engels devoted a separate chapter (1993 [2009], Ch. 4). For these immigrants, living conditions were still worse, but like Carlyle, Engels gave the impression that this was the fault of the Irish themselves, a 'race' who were (his narrative suggests) naturally more brutish, more animalistic (Martin 2004, 95–7). The immigrants were, Engels claimed, 'intemperate, and improvident' (101), possessed of 'brutal habits' (101) and inclined to 'bestial drunkenness' (103); a 'lack of cleanliness' was 'the Irishman's second nature' (103), and 'his crudity … places him little above the savage' (104). As these examples suggest, Engels's emotive language encodes then prevalent racist assumptions – he quotes (101–2) Carlyle's own diatribe against Irish immigrants – which mirror the dualistic divide described by Val Plumwood, whereby marginalised communities or populations are identified with an inferiorised 'nature' (Plumwood 1993 [1997], 43–5). In Engels's account, the Irish formed a separate, still lower form of labouring class, a residuum that the English working class ought itself to fear, because the Irish were prepared to work for even less pay. As levels of pay fell, Engels argued, the related danger was that the English labourer would himself be 'degraded by the presence of Irish competition' (105), a danger which was in any case implicit in slum living conditions so appalling that it was difficult to imagine how anyone could retain her or his humanity.

As this discussion suggests, Engels's account of working-class life in early Victorian Britain offers a graphic account of environmental inequality and its material impacts – even when, as here, Engels's own analysis does not transcend the prejudices of his time. In the terms of environmental justice scholarship, the labouring communities that Engels describes are disadvantaged in several ways. Reduced to the status of 'hands' in the new factories, eking out their lives as atomised individuals, they lack economic leverage and political participation, notwithstanding emerging forms of labour organisation (such as trade unions) and political movements (most notably Chartism). Moreover, they lack cultural recognition and respect. Their segregation constitutes a form of nonrecognition; they are rendered in effect invisible (Schlosberg 2018, 631). Moreover, they are subject to a pattern of *mis*recognition, witness Engels's own description of Irish communities; they are 'routinely maligned or disparaged in stereotypical public and cultural representations' (Schlosberg 2018, 631). Thus, the interaction between different forms of inequality denies the labouring classes even most the basic opportunities to flourish, a denial reflected in the brutal facts of lowered life expectancies.

But what is it to flourish? Engels's description of Victorian working-class life offers a defining early statement of what constitutes environmental inequality, in which we can recognise aspects of all four main theories of environmental justice scholarship (justice as recognition; procedural or participative justice; distributive justice; and justice as capabilities) (Coolsaet 2021, 2–3). In the shock cities newly shaped by industrial modernity, nonrecognition, and misrecognition prevent the kind of participative involvement that might lead to the more equitable distribution of environmental harms and goods, and foster the capabilities that make for a flourishing existence. Yet the question stands: what is it to flourish? If the underworld he described is a 'Hell upon Earth' (65), a phrase so often used by contemporaries to depict life for the industrial classes, what constructive alternate did the young Engels envisage? As I discussed in Chapter 1, Carlyle invoked a Georgic tradition of farming literature in his vision of a better future; as David Fairer (2011) has argued, the Georgic promises a meaningful, embodied existence, in which to work is to live, and the nature of the work is itself a way of relating to and engaging with natural processes and nonhuman natures, perhaps in a way that might foster a sustainable mode of existence. Engels's own response, by contrast, invoked another ancient literary mode or trope: the pastoral.

In his own account of this powerful but problematic literary trope, Terry Gifford (1999) identifies three kinds of pastoral. The first is rooted in a specific literary tradition that began with Greek and Roman poetry about country life (Gifford 1999, 1). The second kind of pastoral is more general, reflecting the contrast between country and city, and celebrating the former at the expense of the latter (Gifford 1999, 2); it becomes widespread in the wake of the Industrial Revolution (Garrard 2012, 37). Writing a few years after Engels, for example, Elizabeth Gaskell opens her own depiction of industrial life, *Mary Barton* (1848), with a pastoral scene (2003, 5–7), complete with 'the delicious sounds of rural life' (5). This, second celebratory form has, however, been critiqued for its idealisations and evasions (Gifford 1999, 2), generating a 'third, pejorative use' of the term 'pastoral' (Gifford 1999, 5), most notably in Marxist critiques (Garrard 2012, 38); this third use of 'pastoral' is often associated with the work of Raymond Williams (Gifford 1999, 9), since it was Williams who identified the way in which the pastoral involves an elegiac backwards glance to a never-never land whose idealised depiction serves 'to cover and to evade the actual and bitter contradictions of the time' (Williams 1973 [1975], 45).

Engels was himself critical of pre-industrial rural existence, and necessarily so: he sided firmly with a dialectical modernity, and the revolutionary promise of communism. To look back was in and of itself regressive (Bonnett 2010, 1). Moreover, and as Aruna Krishnamurthy points out, Engels recognised that the appalling conditions in industrial conurbations were themselves playing an important part in encouraging revolutionary sentiment: as such, cities like Manchester were 'the site of the symptom *and* the cure' (Krishnamurthy 2000, 437). In describing life before the industrial revolution, when 'spinning and weaving' were carried out at home (15), Engels was therefore careful to point out that the workers were no less beholden to a hierarchical social structure, and they were already 'toiling machines in the service of the few aristocrats who had guided history down to that time' (17). 'They were comfortable in their silent vegetation', Engels dismissively observed, but 'intellectually, they were dead' (17); their existence was simply 'not worthy of human beings' (17). Only when the industrial revolution completed the workers' transformation into 'machines pure and simple' (17), 'so forcing them to think and demand a position worthy of man' (17), was it then possible for them to join the historical flow which would lead ineluctably to revolution.

Yet the pastoral exerts a peculiar, perhaps inescapable pull: Lawrence Buell calls it a 'species of cultural equipment western thought has for more than two millennia been unable to do without' (1995, 32). Even as Engels distanced himself from the appealing myth of 'Merry Old England' (34), a myth at once social and ecological, he unintentionally confirmed its appeal. In the working world he described, every working family had a little land, and the leisure to cultivate it. '[P]ermanently settled', wrote Engels, the working man 'had a stake in the country' (16); wrapped up in 'their looms and gardens'

(17), working folk were able to lead a 'passably peaceful existence' (16). As he added, 'their material position was far better than that of their successors' (16); so, perhaps, was their psychological one, as what Marx called 'natural, corporeal, sensuous' beings (Marx 1975 [1992], 389) engaged in 'life-producing life', the 'free conscious activity [that] constitutes the species character of man' (1975 [1992], 328).

Even in Engels's hands, therefore, the pastoral suggests a 'cosily romantic' ideal (Engels 1993 [2009], 17), and it haunts the book, implying an alternative path for society as a whole. Inevitably, that alternative requires a return to the past, rather than another step away from it, and to that extent the pastoral may indeed be regarded as nostalgic and reactionary. But as Buell argues, the pastoral also offers 'a way of designating what has not yet been greatly transformed by pollution' (2001, 5) whilst 'underscoring the importance of the however-modified nonhuman world to the maintenance of life' (2001, 6). Moreover, the pastoral provides a powerful counterpoint to industrial modernity, and 'the violence and excess of techno-transformation' (Buell 2001, 5); it acquires much of its power from the world with it is juxtaposed. This is no less true of Engels's own use of the pastoral, here positioned in contrast to the working lives of the industrial proletariat, and the slums where 'the moral and customs of the good old times are most completely obliterated' (34).

In setting up this contrast, Engels drew on the Gothic, another powerful literary trope. In the Victorian period, Fred Botting writes in relation to the Gothic, 'the wild landscapes of Romantic individualism give way to terrors and horrors that are much closer to home' (1996, 113). The place of the 'gloomy forest or dark labyrinth' is taken by the city, 'industrial, gloomy' (Botting 1996, 114), and itself 'a locus of real horror' (11). Now, Gothic subjects have themselves become 'alienated, divided from themselves', prey to inner impulses as well as external agentialities (Botting 1996, 12). In this modified, urban Gothic, argues Richard Maxwell, the 'Gothic secret' becomes 'entangled with city life', an entanglement that is already apparent in G. W. Reynolds's vast serial *The Mysteries of London* (1844–8), and is 'authoritatively present' in Charles Dickens's novel *Bleak House* (1852–3) (Maxwell 1977, 188) – discussed in more detail in Chapter 4 – where the secret that the novel sets out to expose is, in part, society's complicity in environmental injustice.

These aspects of a modified, Victorian Gothic – horror and monstrosity, alienation, and secrecy (Willis 2012 [2014], 18) – all play a part in Engels's account of life under the conditions created by industrial modernity. His description of Manchester's Old Town (59–65) is one instance. Precipitously positioned on a 'declivitous hillside' (60), and 'threateningly' overlooked by 'the "Poor-Law Bastille" of Manchester' (62), the Old Town is depicted as a 'planless, knotted chaos' (63) in which there is not enough air, and what there is has been 'utterly corrupted' (65) by the filth in which the courts and alleys are steeped. In contrast to the brilliantly-lit streets frequented by the 'money aristocracy' (58), this infernal region is literally and metaphorically fallen: Engels repeatedly emphasises its ruinous, tumbledown nature, and on three occasions describes it as a chaos. It is a carefully chosen word. Most obviously, 'chaos' captures the unplanned and disorderly nature of the slums, which (like the 'rookeries' of Dickens' *Bleak House*) have sprung from already ruinous housing stock; they are, in simple terms, the consequence of the factory's sharply increased demand for labour, and the factory owner's indifference to the social and ecological cost (65–6). This much forms part of Engels's carefully argued case against industrial modernity. But the word's additional meanings are more emotive, and they relate directly to the Gothic atmosphere that Engels also establishes within the book: to speak of a 'chaos' is to describe a 'gulf, chasm, or abyss', reinforcing the reader's sense of the Old Town as a dark and fallen world, yet the word carries a further meaning, as '"the formless void" of primordial matter' (Little, Fowler, and Coulson 1983 [1988], 314). Like the opening description of London in *Bleak House* – of streets so mired in mud it seems 'as if the waters had but newly retired from the face of the earth' (Dickens 1852–3 [2003], 13) – this meaning suggests a deliquescent, entropic world: the Old Town is reverting to an elemental state, or perhaps something worse, a suggestion reinforced by Engels's insistent, repeated use of the word 'filth' to describe conditions in the alleys and courts. Like the dust of Dickens' last novel, *Our Mutual Friend*, discussed in the Introduction, filth is Engels's code for an amalgam of every kind of waste, including decaying animal matter and human excrement. This is the matter into which the slums are sinking – or perhaps the filth, now animate, is itself rising up to swallow everything around it – and this becomes, in turn, another source of the horror (63) we are made to feel as Engels guides us through these forgotten, overlooked streets.

This is not just a shut-off, a secluded world (63), into which the well-to-do avoid looking; crucially, it is a labyrinth – an 'ill-built, ill-kept labyrinth' (63) – a maze, an entanglement, 'intricate and tortuous' (Little, Fowler, and Coulson 1983 [1988], 1166), corresponding exactly to the Gothic trope's archetypes of 'entrapment and decay' (Willis 2012 [2014], 19).[4] This is Engels's final point, the final element in the transformation of factual account into Gothicised narrative – that these slum quarters are not only unknown, but in a sense unknowable. Whosoever takes a wrong turn here 'is lost; he wanders from one court to another, turns countless corners, passes nothing but narrow, filthy nooks and alleys, until after a few minutes he has lost all clue, and knows not whither to turn' (62–3). The slum marks the limit of understanding, which in turn requires the 'Virgilian mode' (Buell 2001, 43) that Engels offered his readers, as he 'penetrates [the] chaos' (Engels 1943 [2009], 63). Naturally, this tour of the city's slums – a place blanked by polite society – comes with a map (61); just as inevitably that map is partial and incomplete; courts and alleys that defy description now also refuse to be named or fully charted.

What Engels's Gothicised depiction gives the reader, therefore, is indeed a 'Hell upon Earth' (65), complete with a bubbling, 'coal-black, foul-smelling stream' (62). It is a place of literal gloom and metaphorical darkness, where enlightenment has failed, and rationality is giving way to irrationality. Engels's Old Town is a place, in other words, of (revolutionary) risk, which Beck describes in terms that themselves signal its shadowy nature, its abyssal identity. 'Threats from society are bringing about a kind of new "shadow kingdom"', Beck wrote, 'comparable to the realm of the gods and demons in antiquity, which is hidden behind the visible world and threatens human life on this Earth' (1992 [2012], 72). This heightened sense of risk also suggests an important link to new materialist thinking. As I outlined in the Introduction, new materialist theories problematise the 'environment' in environmental justice, by highlighting both its own agentiality – its own 'thing-power' (Bennett 2010, xvii) – and the constitutive entanglements between people and place. As Jane Bennett stresses, things possess their own vibrancy, a vibrancy that eludes 'the violent human will to dominate and control' (2010, xvii). As she adds, a new materialist approach emphasises (and may 'even overemphasise') 'the agentic contributions of nonhuman forces' as they operate 'in nature, in [and on] the human body, and in human artefacts' (Bennett 2010, xvi).

At one, very obvious level, this sense of matter's unpredictability maps readily onto Engels's own account of factory life, where workers were daily exposed to the 'independence and aliveness' (Bennett 2010, xvi) of machines that were never intended to mutilate or maim, but did so with appalling regularity (Engels 1993 [2009], 174). Here, entanglement is a literal source of life-altering or life-threatening risk: this is the impact of industrial modernity at its most visceral and self-evident. But 'thing-power' also operates in more insidious ways, as Morton has argued in relation to hyperobjects, which may intrude on human affairs, or hover at the very limits of human comprehension; indeed, industrial modernity may itself be thought of as a hyperobject, a 'vampire-like' assemblage of other entities, 'sucking away' at both humans and nonhumans (Morton 2013, 5), whilst systematically reshaping the earth itself (4).[5] The related challenge, as Morton contends, is that (by definition) the hyperobject 'is not a function of our knowledge' (2013, 2); like Beck's (re)defined notion of risk, the hyperobject marks the limits of human understanding, and by extension, our ability to grasp it using existing forms of language.

This was no less a challenge for Engels, wrestling with the obvious iniquities of environmental classism, yet nonetheless aware of (if not always able to articulate) the wider effects of industrial modernity as they encompassed the natural and nonhuman world. Take, for example, Engels's use of that semantically slippery word 'nature', 'perhaps the most complex word in the [English] language' (Williams 1983 [1988], 219). Engels uses the word in several different ways, of which two are relevant to this discussion. First, he refers to (inherent) human nature (36, 37, 77, 301), the nature, that is, which industrial modernity is debasing. Second, he refers to 'Nature', with an initial capital letter to indicate that what is being described is distinct from human nature; it is the 'Nature' that is set apart by Plumwood's dualistic divide. This 'Nature' itself takes several forms: there are 'the laws of Nature' (246) and 'the forces of Nature' (33, 54, 93), uses that imply both its deterministic and its agential significance, but there is another sense in which Engels uses the term, which suggests something of the Romantic view of it a source of solace and comfort, of 'Nature' as 'a lost world of peace and companionship, of healthy bodies and quiet minds' (Houghton 1957, 79), quite remote from the modern, industrial conurbation. It is in this spirit that Engels uses the word when he refers to 'the enjoyment of Nature' now denied to the working class (130), and to the 'working man [who] never gets the slightest glimpse of Nature in his large town with his long working hours' (246). It is there again when Engels prefaces his accounts of Manchester's urban squalor with a brief mention of the 'beautiful hill country' and 'charming green valleys' that surround it (53). And it is there, of course, in his opening sketch of a society before its transformation by industrial capitalism, as the benign accompaniment to a 'Merry Old England' (34).

As a succession of writers has pointed out, this view of England's 'Nature' obscures the long history of human involvement in its construction: as Williams observed, the English hedgerow has been 'planted and tended', and '[a] considerable part of what we call natural landscape has the same kind of history' (1983 [1988], 78; see also Taylor 2016, 4). In this sense, the term 'Nature' gets in the way of our understanding of the interrelationships between people and places; to be 'properly ecological', Morton has argued, society must do away with it (2007 [2009], 1). But if not the word 'nature', then what? We can use the term 'naturecultures' or 'naturalcultural', as do Donna Haraway (2008, 16) and Karen Barad (2007, 32), or 'nature-culture', as Bruno Latour has done (1993, 7), but this conflation of one with the other serves to obscure the impact of human on nonhuman, an impact which grew markedly more intense and destructive with the advent of industrial modernity. As Jesse Oak Taylor argues, there is still a great deal of difference 'between a cultivated landscape and a decimated one' (2016, 6), and we need a way in which to register that distinction.

With this in mind, Taylor proposes a new term, the *abnatural*, on which I draw throughout this book. As Taylor explains, the term denotes both 'nature's absence and its uncanny persistence', as it continues to 'adapt, mutate, migrate, and evolve, even under artificial conditions', 'always eluding definition' (Taylor 2016, 5). A manufactured or built environment (all bricks and mortar) may be abnatural; so are the weeds which, against all odds, have found ways in which to colonise the tallest chimneys. But the abnatural also describes the experience of living amidst these conditions; it 'characterises the felt presence of an absence, a partial glimpse of an open whole' (Taylor 2016, 6). This is, Taylor argues, an expression that reflects 'the experience of dwelling in flux' (Taylor 2016, 6): one can refer to the abnatural, but not to abnature; the abnatural is always dynamic and elusive, and never settles into a single, stable state.

If we bring this sense of the abnatural to bear on Engels's account of Manchester slums, we can begin to see some of the ways in which his account of slum life itself registers the wider dimensions of 'ecological trauma' (Taylor 2016, 5). Nothing green seems to grow in the world Engels describes, but he does note how the now 'coal-black, foul-smelling' Irk deposits 'the most disgusting, blackish-green, slime pools [along the] bank, from the depths of which bubbles of miasmatic gas constantly arise and give forth a stench unendurable even on the bridge 40 or 50 feet above the surface of the stream' (62). This may not be a Romanticised 'Nature', as the Victorians preferred to think of it, but it is abnatural, in the threefold sense that Taylor intends for the term: it marks not only an absence (of an unpolluted ecosystem), but also a persistence, of nature or of natural processes in a new and no less active form; in turn, it offers a glimpse of a whole, denoted not just by the obvious inference that the Irk will carry these effluents and pollutants downstream, into the world beyond this underworld, but by Engels's reference to 'bubbles of miasmatic gas'. It is a significant reference. Substantial interest has been taken in novels like *Bleak House*, with its unifying image of a London smog that anticipates our own, late-modern anxieties about a changed climate (Taylor 2016, Ch. 1); Engels, by contrast, focuses on a different kind of airborne problem, characterised not by smoke, but by smell. For Engels, as for so many of his contemporaries, the problem was not simply that the squalor of the slums created 'a stench [as in Manchester's Old Town] which alone would make it impossible for a human being in any degree civilised to live in such a district' (63); that stench was a symptom of a deeper and much more disturbing transformation, as the atmosphere itself was thrust into the foreground as something at once threatened and threatening. It is no longer life-supporting or life-enhancing, as it might still be, Engels noted, for those who live in the country, or for those members of the bourgeoisie fortunate enough to live on the city's hilly outskirts; it is itself alive, and menacing, even lethal.

> All putrefying vegetable and animal substances give off gases decidedly injurious to health, and if these gases have no free way of escape, they inevitably poison the atmosphere. The filth and stagnant pools of the working-people's quarters in the great cities have, therefore, the worst effect upon the public health, because they produce precisely those gases which engender disease; so, too, the exhalations from contaminated streams.
>
> (108)

As these observations highlight, Engels believed in a miasmatic theory of disease transmission, a then widely held view that rotting organic matter engendered infected or 'bad air' (Halliday 2001, 1469): in the words of Edwin Chadwick, architect of the Poor Law, '[a]ll smell is, if it be intense, immediate acute disease' (1846; qtd in Halliday 2001, 1469). For the slum-dwellers, Engels wrote, it was a problem of too little air, shared between and depleted by the many who crowded these districts (107), of air 'imprisoned' in courts (67) where there was little or no ventilation (another of the words that Engels repeatedly uses to describe the problems of the slums) (67–8), but above all of an 'atmosphere' that had been 'utterly corrupted' (65), 'poisoned by the effluvia from … masses of refuse, offal and sickening filth' (72). '[E]verything combines to poison the air' (108), he wrote.

In the cities created by industrialisation, in other words, and in the slums that the economy's iron laws create, the air has ceased to occupy its usual role as a taken-for-granted backdrop to human activity. The air itself has come alive, in an abnatural form triggered (but no longer controllable) by human activity. (The parallel with climate change is an obvious one.) This has immediate and unavoidable material consequences. The human body does not itself possess 'independently determinate boundaries and properties' (Barad 2007, 33); the body is open, porous (Alaimo 2010, 2–3). As Engels insisted, slum-dwellers cannot avoid the consequences of an atmosphere that is now both agential and toxic – harmful, insidious, virulent – and which manifests itself in epidemics that spread through the slums with periodic inevitability. As Engels also pointed out, the physical segregation of the poor is not itself sufficient to confine those epidemics to the slums. If disease is airborne – if the air itself has changed its character, and become a kind of living poison – then the well-to-do may be at risk, even as they retreat to the edges of towns or occupy the new suburbs. It was as these epidemics approached, wrote Engels, that '[p]eople remembered the unwholesome dwellings of the poor, and trembled before the certainty that each of these slums would become a centre for the plague, whence it would spread desolation in all directions through the houses of the propertied class' (76); and then, Engels (gleefully) added, 'a universal terror seized the bourgeoisie of the city' (76). '[*S*]*mog is democratic*', Beck famously declared (italics in the original; 1992, 36); in the toxic risk society of Manchester, suggested Engels, it is a miasmatic atmosphere that knows no bounds (or boundaries).

There was, however, a second and less literal sense in which this abnatural, 'abnormal atmosphere' (Engels, 1993 [2009], 107) might rebound on the better-off – a second, and a revolutionary risk to comfortable bourgeois lives. As Engels stressed, the stench of the slums was an indictment of bourgeois indifference – an indifference he characterised as social murder – and it formed part of a wider pattern of environmental inequality that extended far beyond the localised instances of injustice in Manchester's Old Town. 'The manner in which the great multitude of the poor is treated by society today is revolting', Engels explained:

> They are drawn into the large cities where they breathe a poorer atmosphere than in the country; they are relegated to districts which, by reason of the method of construction, are worse ventilated than any others; they are deprived of all means of cleanliness, of water itself, since pipes are laid only when paid for, and the rivers so polluted that they are useless for such purposes; they are obliged to throw all offal and garbage, all dirty water, often all disgusting drainage and excrement into the streets, being without other means of disposing of them; they are thus compelled to infect the region of their own dwellings.
>
> (108)

'All conceivable evils are heaped upon the heads of the poor', Engels concluded (108). What might the long-term effect be, of environmental injustice on so large a scale? If we conceive of those 'evils' as a form of 'incremental and accretive … "slow violence"', as Rob Nixon suggests (2011, 2), how might this affect its victims? It was not just a question of the debilitating physical effects of these conditions on the labouring classes, and of the moral impact on a people forced to exist (or subsist) in the most squalid conditions. Engels (1993 [2009], 128–9) quoted Carlyle, and Carlyle's essay on *Chartism* (1839 [1897]), in which Carlyle explicitly describes the suffering of the working classes in terms of (in)justice, and its likely, 'fatal' result in revolution (Carlyle 1839 [1897], Vol 16, 233).[6] 'This rage', wrote Engels, is 'proof that the workers feel the inhumanity of their position, that they will refuse to be degraded to the level of brutes, and that they will one day free themselves from servitude to the bourgeoisie' (129).

But there was another possibility, about which Victorians also speculated: that there would no revolution; that the workers would not rise up, but instead succumb, and find themselves transformed into the 'brutes' to which Engels referred. Carlyle invoked 'Nature's order' (Carlyle 1839 [1897], Vol 16, 232; qtd in Engels, 128), as though 'nature' was static and immutable, a standard against which every behaviour could be measured, but the reality of a pestilential, miasmatic atmosphere suggested that there was nothing that could not be corrupted, disturbed, or changed. What if people were themselves corrupted, their own inherent natures somehow affected? In describing the impact of work in the mills and factories, Engels stressed not only the injuries and diseases to which it gave rise, but also the deformations it induced, caused (for example) by 'frequent stooping and … bending' (173). These deformities were permanent (Clark and Foster 2006, 382); they gave rise, Elizabeth Gaskell remarked in her novel *Mary Barton* (1848), to 'a through specimen of a Manchester man', 'stunted', 'wan, colourless' (2003, 7). Was it perhaps possible that, under the pressure of conditions such as these, the workers might themselves be transformed, become *abnatural*? It was a disturbing thought, widespread in the later Victorian period, when there was much talk of 'degeneration', the title of Max Nordau's influential tract (*Entartung*, 1892–3). Notably, H. G. Wells wove the idea into *The Time Machine* (1895): as his time-traveller discovers, the working classes have evolved into the Morlocks, a race of predatory, bestial creatures who periodically emerge from their underworld to predate on the Eloi, descendants of the affluent suburbanites who once pretended ignorance of the slum quarters on which their wealth was built. Wells's fiction is, of course, speculative, but as a thought experiment, the idea of this evolutionary schism is all the more disturbing because the seeds of it are already there, in Engels's descriptions of environmental inequalities so profound that Engels thinks it inconceivable that this kind of transformation is not already underway. 'In a word', he wrote,

> we must confess that in the working-men's dwellings of Manchester … only a physically degenerate race, robbed of all humanity, degraded, reduced morally and physically to bestiality, could feel comfortable and at home. And I am not alone in making this assertion.
>
> (75)

*

Engels's *Condition of the Working Class* remains the defining account of environmental injustice in Victorian Britain, an injustice that was, as Engels insisted, neither localised nor temporary. It was rather a permanent and enduring condition of industrial modernity, of capitalism in the ascendant, and it was felt most acutely by the working classes, whose living and working conditions justify Pellow's argument that whole populations may come to be seen as 'expendable' (Pellow 2018, 17). This was environmental classism, a naturalcultural interchange or 'intra-action' whereby (as Beck has subsequently argued) human impacts on the nonhuman world rebound on society, altering social relations, perhaps permanently. What the Victorians thought of as 'Nature' came to seem, as Taylor has argued, *abnatural*: not simply absent (although absence can itself be constitutive), but present in ways that do 'not comply with our ideas of nature' (2016, 5); present, that is, in disturbing new ways, sometimes insistent, sometimes elusive. As new materialist theories emphasise, matter has agency: it is not the pliable product of human society, but is itself dynamic; it intra-acts with (and impacts on, constrains, and constructs) that society; and it was this recognition – that an abnatural, agential reality was itself making changes to human social structures – that Engels depicted with such literary vigour.

What Engels described, Marx was already analysing. The conditions created by industrial modernity, Marx insisted, created an atomised, alienated existence characterised by what he called *Entfremdung*, or estrangement of the self, a theory he began to develop in the 'Economic and Philosophical Manuscripts' (1844) written at the same time as Engels's *Condition of the Working Class*. Industrial capitalism, wrote Marx, 'estranges nature from man [and] man from himself, from his own active function, from his vital activity' (Marx 1975 [1992], 328).[7] 'What requires explanation is not the *unity* of living and active human beings with the natural, inorganic conditions of their metabolism with nature', he added in his later notes on *Pre-Capitalist Economic Formations* (italics in the original; written 1857–8; Marx 1965, 85).[8] 'What we must explain is the *separation* of these inorganic conditions of human existence from this active existence, a separation which is only fully completed in the relationship between wage-labour and capital' (italics in the original; Marx 1965, 87). For Marx, as for Engels, revolution would necessarily entail 'the overthrow of alienated nature and alienated labour', and hence a (re)turn 'to the wealth of human-natural interconnections' (Foster and Burkett 2016 [2017], 53). As John Bellamy Foster and Paul Burkett point out,

> [t]he valuing of nature in this perspective was not that of an abstract, distant contemplation, reflecting the reality of alienation; but rather something that was to be realised in a real, sensuous way, in the form of an active, material dependence, within a higher form of society.
>
> (2016 [2017], 54)

The wider question is whether, nearly two centuries later, we are any closer to the kind of environmentally just world this implies. '[D]espite ongoing and strenuous efforts', Karen Bell has argued (2018), 'environmental justice seems as elusive as it did when the term was first coined in the 1980s' (2018, 543). As she adds, 'I have come to the conclusion that this is because environmental justice is consistently undermined by capitalism with its drives towards inequality and irrationality' (Bell 2018, 543). Engels would undoubtedly have agreed. In *The Condition of the Working Class in England*, Engels points the finger for the environmental iniquities generated by industrial capitalism directly at 'the money-greed of the bourgeoisie' (159). The bourgeoisie was, he wrote, 'incurably debased by selfishness, [and] incapable of progress' (280); it was 'already too late to bring about a peaceful solution' (301). 'The revolution must come', he insisted (301). 'Prophecy is nowhere so easy as in England' (301).

That belief in the certainty of revolution lends Engels's account its optimistic note: in his view, change was inevitable, and that change would liberate the working classes and enable them to flourish. Writing in the 1840s – the Hungry Forties, as they became known – Engels had reason to be hopeful. Like Carlyle, he was responding to a remarkable surge of political activism that emerged from within the working classes in direct response to the conditions created by industrial capitalism. This was Chartism, a movement to which Engels also contributed his skills as a reporter. Like Carlyle, however, Engels was writing as a (middle-class) outsider, with his own beliefs and biases about the right direction for society; Carlyle's innate conservatism was reflected in his attempt to transpose an older, aristocratic hierarchy on to modern, industrialised society; Engels's youthful radicalism was reflected in his willingness to assume the inevitably of a revolutionary outcome.

How, then, did the working classes portray themselves? The next chapter engages with a remarkable body of work produced by Chartist writers in the early Victorian period. For Chartists, the dire nature of living and working conditions was only the most immediate problem with which they wrestled. Increasingly, working-class lives were dictated by the free play of industrial capitalism, unconstrained by state intervention or regulation; the working class was itself both economically disempowered (lacking any control over the means of production) and politically disenfranchised (their exclusion from the terms of the Great Reform Act of 1832 was itself a spur to the movement's emergence). Nor was this the only disadvantage under which the working classes laboured. Concerned middle-class critics and commentators were quick to condemn the impact of industrial modernity on the working classes, and in so doing, highlight the abhorrent nature of slum life. As I have highlighted, however, they also drew a connection between environmental degradation, and what they perceived to be working-class demoralisation, even 'degeneracy'; and if the working classes were 'degenerate', they were not and could not be morally or politically considerable. For the Chartists, therefore, it was vital that these misrepresentations – these manifestations of misrecognition – be contested, challenged, inverted. As I discuss, the result was a substantial body of literature spanning prose and poetry, variously concerned to establish working-class legitimacy and considerability, whilst also wrestling with the obvious question to which environmental inequality gave rise: what might a just society look like?

*

Notes

1 The copy text for this chapter is the Oxford University Press edition (1993) of *The Condition of the Working Class*, to which all in-text citations refer.

2 Class is often understood in terms of income, but as Karen Bell points out, it also encompasses a social and cultural dimension, and both relational and gradational aspects (2020, 27–30): in her own discussion of environmental classism, she 'includes the material distribution of wealth and income but also "recognition", status and valuing' (2020, 30). As Bell stresses, this (re)definition of Marxist conceptions of class does not rule out (indeed explicitly encompasses) intersectional links to other forms of oppression (for example, 'race, disability, LGBTQ+ identity, [and] age' (Bell 2020, 18). Clearly, Bell's definition is an inclusive and fluid one. Engels himself is more narrowly concerned with class in its economic sense, for reasons that my own analysis underlines, but he is alert to the ways in which cultural and discursive constructions shape and influence status, not least in relation to the dualistic divide that Val Plumwood has highlighted (1993 [1997]). In turn, that divide has been variously used within society to inferiorise and 'other' a range of marginalised groups, including those defined in terms of gender, race, and sexuality.

3 Engels's argument is in sharp contradistinction to the claim – made then and since – that these conditions were created by the accident of insufficient and inadequate housing stock; see, for example, T. S. Ashton's dismissal of industrial capitalism as a cause (1954 [1963], 49). The fact remains that factory owners were the beneficiaries of a plentiful supply of labour whose living conditions were not their (economic or social) concern. Meanwhile, speculators saw no incentive in building better dwellings for the very poor when money could still be extracted from them by renting out and subdividing the most ruinous premises (Stedman Jones 1971 [2013], 173). Moreover, there was nothing accidental about the siting of 'tanneries, bonemills, and gasworks' in the middle of working-class districts (Engels 1993 [2009], 62); as I noted in the Introduction, 'nuisance' industries could be imposed on the working poor because they had neither the (political) will nor the (financial) wherewithal to resist. Engels's point is, of course, that this sense of powerlessness itself stimulates revolutionary resistance.

4 There is a further sense to the word, which Engels may or may not have intended: 'labyrinth' also suggests the mythological lair of the monstrous Minotaur. Elsewhere, Engels refers to the slums of Old Town as 'cattle-sheds for human beings' (63), begging the question: if humans were treated as animals, might they not come to resemble beasts? That logic was already inherent in Engels' racial categorisation of the working-class Irish community, a prejudice shared with Carlyle and others, but as I discuss in Chapter 3, this discourse of environmental determinism – with its negative constructions of the labouring poor – became a significant issue, with which Chartist writers (not unproblematically) grappled.

5 Thus, Morton argues that the Anthropocene dates from 1784, when James Watt's improvements transformed the steam engine into 'an all-purpose' device, and so 'precipitated the industrial age' (Morton 2013, 4). Others, such as Jason W. Moore, point to 'capitalism's early-modern origins, and its extraordinary reshaping of global natures long before the steam engine', and on this basis, argue that the term Anthropocene should be replaced by 'Capitalocene' (Moore 2017, 3).

6 Engels (128–9) quotes as follows. 'Injustice, infidelity to truth and fact and Nature's order, being properly the one evil under the sun, and the feeling of injustice the one intolerable pain under the sun, our grand question as to the condition of these working men would be: Is it just? And first of all, what belief have they themselves formed about the justice of it? The words they promulgate are notable by way of answer; their actions are still more notable ... Revolt, sullen revengeful humour of revolt against the upper classes, decreasing respect for what their temporal superiors command, decreasing faith for what their spiritual superiors teach, is more and more the universal spirit of the lower classes. Such spirit may be blamed, may be vindicated; but all men must recognise it as extant there, all may know that it is mournful, that unless altered it will be fatal'. The original source can be found in Carlyle 1839 [1897], Vol 16, 232–3.

7 Under the conditions created by capitalism, Marx argued, workers and labourers no longer have control over what they produce, or how they produce it; even their own labour (as factory systems of discipline demonstrated) is no longer theirs to control, simply to sell. Just as the worker's labour itself becomes an alien object, 'and begins to confront him as an autonomous power', so 'the life which he has bestowed on the object confronts him as hostile and alien' (the more of himself he gives to the object, the more life it possesses) (Marx 1975, 324). Workers are thereby alienated from themselves, and from the products of their own labour, a process of estrangement that is itself integral to the dualistic logic of colonisation that Plumwood identifies in Western thought.

8 *Pre-Capitalist Economic Formations* formed part of the notes Marx wrote in preparation for *A Contribution to the Critique of Political Economy*, published

in 1859. These notes were eventually published in Russian between 1939–41 and in German in 1953, when they took the name by which they are now known: the *Grundrisse der Kritik der Politischen Ökonomie*, or, for short, the *Grundrisse* ('Outline').

Works cited

Alaimo, Stacey. 2010. *Bodily Natures: Science, Environment, and the Material Self*. Bloomington, IN: Indiana University Press.

Ashton, T. S. 1954 [1963]. 'The Treatment of Capitalism by Historians'. *Capitalism and the Historians*, edited by F. A. Hayek, 31–61. Chicago, IL: University of Chicago Press.

Barad, Karen. 2007. *Meeting the Universe Halfway: Quantum Physics and the Entanglement of Matter and Meaning*. London: Duke University Press.

Beck, Ulrich. 1992 [2012]. *Risk Society: Towards a New Modernity*, translated by Mark Ritter. London: Sage.

Bell, Karen. 2018. 'The Dawn of Environmental Justice? The Record of Left and Socialist Governance in Central and South America'. *The Routledge Handbook of Environmental Justice*, edited by Ryan Holifield, Jayajit Chakraborty, and Gordon Walker, 543–55. Abingdon: Routledge.

Bell, Karen. 2020. *Working-Class Environmentalism: An Agenda for a Just and Fair Transition to Sustainability*. Cham, Switzerland: Palgrave Macmillan.

Bennett, Jane. 2010. *Vibrant Matter: A Political Ecology of Things*. London: Duke University Press.

Bonnett, Alastair. 2010. *Left in the Past: Radicalism and the Politics of Nostalgia*. London: Continuum.

Botting, Fred. 1996. *Gothic*. London: Routledge.

Buell, Lawrence. 1995. *The Environmental Imagination: Thoreau, Nature Writing, and the Formation of American Culture*. London: Belknapp Press.

Buell, Lawrence. 2001. *Writing for an Endangered World: Literature, Culture, and Environment in the U.S. and Beyond*. London: Belknapp Press.

Cannadine, David. 2017. *Victorious Century: The United Kingdom, 1800–1906*. London: Penguin.

Carlyle, Thomas. 1839 [1897]. *Chartism. The Works of Thomas Carlyle, Volume Sixteen*, 204–85. New York, NY: Peter Fenelon Collier. Available at archive.org, https://archive.org/details/worksofthomascar0016carl/page/n7/mode/2up?view=theater. Accessed 23 November 2022.

Clark, Brett, and John Bellamy Foster. 2006. 'The Environmental Conditions of the Working Class: An Introduction to Selections from Frederick Engels's *The Condition of the Working Class in England in 1844*'. *Organization & Environment* 19 (3): 375–88. Available at jstor.org, http://www.jstor.org/stable/26162420. Accessed 22 November 2022.

Coolsaet, Brendan. 2021. 'Introduction'. *Environmental Justice: Key Issues*, edited by Brendan Coolsaet, 1–5. London: Routledge.

Dickens, Charles. 1852–3 [2003]. *Bleak House*, edited and introduced by Nicola Bradbury, preface by Terry Eagleton. London: Penguin.

Disraeli, Benjamin. 1844 [1982]. *Coningsby, or The New Generation*, edited by Sheila M. Smith. Oxford: Oxford University Press.

Engels, Friedrich. 1993 [2009]. *The Condition of the Working Class in England*, edited and introduced by David McLellan. Oxford: Oxford University Press.

Faber, Daniel. 2018. 'The political economy of environmental injustice'. *The Routledge Handbook of Environmental Justice*, edited by Ryan Holifield, Jayajit Chakraborty, and Gordon Walker, 61–73. Abingdon: Routledge.

Fairer, David. 2011. "'Where Fuming Trees Refresh the Thirsty Air': The World of Eco-Georgic'. *Studies in Eighteenth-Century Culture* 40: 201–18. doi:10.1353/sec.2011.0006

Foster, John Bellamy, and Paul Burkett. 2016 [2017]. *Marx and the Earth: An Anti-Critique*. Chicago, IL: Haymarket.

Garrard, Greg. 2012. *Ecocriticism*. 2nd edn. Abingdon: Routledge.

Gaskell, Elizabeth. 2003. *Mary Barton: A Tale of Manchester Life*, edited by Macdonald Daly. London: Penguin.

Gifford, Terry. 1999. *Pastoral*. London: Routledge.

Gill, C. I. C. 1947. 'Byssinosis in the Cotton Trade'. *British Journal of Industrial Medicine* 4 (1): 48–55. Available at jstor.org, www.jstor.org/stable/27720626. Accessed 9 February 2022.

Gratton, Peter. 2014 [2015]. *Speculative Realism: Problems and Prospects*. London: Bloomsbury Academic.

Halliday, Stephen. 2001. 'Death and miasma in Victorian London: An obstinate belief'. *British Medical Journal (BMJ)* 323: 1469–71. Available at bmj.com, https://doi.org/10.1136/bmj.323.7327.1469. Accessed 2 December 2022.

Haraway, Donna J. 2008. *When Species Meet*. London: University of Minnesota Press.

Hilton, Boyd. 2006. *A Mad, Bad, and Dangerous People? England 1783–1846*. Oxford: Clarendon Press.

Hobsbawm, Eric. 1975 [1995]. *The Age of Revolution*: 1789–1848. London: Weidenfeld & Nicolson.

Houghton, Walter E. 1957. *The Victorian Frame of Mind, 1830–1870*. London: Yale University Press.

Krishnamurthy, Aruna. 2000. "'More than Abstract Knowledge': Friedrich Engels in Industrial Manchester'. *Victorian Literature and Culture* 28 (2): 427–48. Available at jstor.org, www.jstor.org/stable/25058528. Accessed 22 November 2022.

Latour, Bruno. 1993. *We Have Never Been Modern*, translated by Catherine Porter. Cambridge, MA: Harvard University Press.

Little, William, H. W. Fowler, and Jessie Coulson. 1983 [1988]. *The Shorter Oxford English Dictionary*, Volume I. 3rd ed. London: Guild.

Martin, Amy E. 2004. 'Blood Transfusions: Constructions of Irish Racial Difference, the English Working Class, and Revolutionary Possibility in the Work of Carlyle and Engels'. *Victorian Literature and Culture* 32 (1): 83–102. Available at jstor.org, http://www.jstor.org/stable/25058653. Accessed 22 November 2022.

Marx, Karl. 1965. *Pre-Capitalist Economic Formations*, translated by Jack Cohen, edited and introduced by E. J. Hobsbawm. New York: International Publishers.

Marx, Karl. 1975 [1992]. 'Economic and Philosophical Manuscripts' (1844). *Early Writings*, translated by Rodney Livingstone and Gregor Benton, 279–400. London: Penguin.

Maxwell, Richard. C. 1977. 'G. M. Reynolds, Dickens, and the Mysteries of London'. *Nineteenth-Century Fiction* 32 (2): 188–213. Available at jstor.org, https://doi.org/10.2307/2933188. Accessed 10 January 2023.

Moore, Jason W. 2017. 'The Capitalocene, Part I: On the Nature and Origins of our Ecological Crisis'. *The Journal of Peasant Studies* 44 (3): 1–37. doi: 10.1080/03066150.2016.1235036

Morton, Timothy. 2007 [2009]. *Ecology Without Nature: Rethinking Environmental Aesthetics*. London: Harvard University Press.

Morton, Timothy. 2013. *Hyperobjects: Philosophy and Ecology after the End of the World*. London: University of Minnesota Press.

Morton, Timothy. 2014. 'Victorian Hyperobjects'. *Nineteenth-Century Contexts* 36 (5): 489–500. doi: 10.1080/08905495.2014.974940

Nixon, Rob. 2011. *Slow Violence and the Environmentalism of the Poor*. Cambridge, MA: Harvard University Press.

Pellow, David N. 2000. 'Environmental Inequality Formation: Towards a Theory of Environmental Injustice'. *American Behavioural Scientist* 43 (4): 581–601. doi:10.1177/0002764200043004003

Pellow, David Naguib. 2018. *What is Critical Environmental Justice?* Cambridge: Polity.

Piketty, Thomas. 2014 [2017]. *Capital in the Twenty-First Century*, translated by Arthur Goldhammer. Cambridge, MA: Belknap Press. Originally published in 2013.

Plumwood, Val. 1993 [1997]. *Feminism and the Mastery of Nature*. London: Routledge.

Schlosberg, David. 2018. 'Environmental (In)justice'. *Companion to Environmental Studies*, edited by Noel Castree, Mike Hulme, and James D. Proctor, 629–34. London: Routledge.

Sherwood, John M. 1985. 'Engels, Marx, Malthus, and the Machine'. *The American Historical Review* 90 (4): 837–65. Available at jstor.org, https://doi.org/10.2307/1858842. Accessed 24 November 2022.

Stedman Jones, Gareth. 1971 [2013]. *Outcast London: A Study in the Relationship Between Classes in Victorian Society*. London: Verso.

Taylor, Jesse Oak. 2016. *The Sky of Our Manufacture: The London Fog in British Fiction from Dickens to Woolf*. London: University of Virginia Press.

Tocqueville, Alexis de. 1835 [1958]. *Journeys to England and Ireland*, translated by George Lawrence and K. P. Mayer, edited by J. P. Mayer. New Haven, CO: Yale University Press.

Waldron, H. A. 1983. 'A brief history of scrotal cancer'. *British Journal of Industrial Medicine* 40 (4): 390–401. doi:10.1136/oem.40.4.390

Williams, Raymond. 1973 [1975]. *The Country and the City*. New York: Oxford University Press.

Williams, Raymond. 1983 [1988]. *Keywords: A vocabulary of culture and society*. London: Fontana Press.

Willis, Martin. 2012 [2014]. 'Victorian Realism and the Gothic'. *The Victorian Gothic*, edited by Andrew Smith and William Hughes, 15–28. Edinburgh: Edinburgh University Press.

3 Environmental determinism and the Chartist counter-narrative

This chapter examines the way in which, in the early Victorian period, working-class activists sought to re-write the contemporary narrative of environmental determinism, with its assumption that physical deprivation and environmental degradation inevitably led to working-class 'degeneration'. As Lawrence Buell has pointed out, the question of environmental determinism took on a new importance in the nineteenth century, as industrial modernity created a toxified and segregated society, riven by environmental injustice (2001, 130–1); increasingly, it was assumed that the brutalised existence to which Friedrich Engels drew attention would create a brutish people, who might deserve a patriarchal state's care and consideration, but not a role in deciding their own future. This was what the historian Boyd Hilton describes as the 'great phantasmagoria' of a 'mad, bad, and dangerous people, an infectious disease threatening to destroy civilisation' (2006, 581).

In the early Victorian period, the challenge of contesting this pejorative construction was taken up by the Chartists, who led 'the first (and arguably still the greatest) mass political movement in industrial Britain' (Chase 2007, 7). The Chartists sought to advance their cause not only by taking to the streets and presenting petitions, but by advancing new narratives about their nature, needs, and aspirations. The working classes were, the Chartists insisted, more than the sum of slum conditions or the product of factory life: they too possessed agency.[1] But as their own lived experience underlined, the human body is always already entangled in a material world, and subject to it: as Richard Lewontin and Richard Levins insist, 'your body knows your class position no matter how well you have been taught to deny it' (2007, 306). For the Chartist writer, therefore, the challenge was one of (re)presenting working-class identity without either reproducing the logic of bourgeois contemporaries, who assumed that demoralisation and depravation followed deprivation, or simply enacting its crude reversal, by denying any kind of environmental influence.

With these concerns in mind, this chapter explores a range of Chartist literature, most of it now forgotten. It begins with 'A Simple Story' (1840), an anonymous narrative which engages head-on with assumptions about working-class 'depravity', and counters them with an account of working-class communities of care that continued to exist even (or perhaps especially)

DOI: 10.4324/9780367821609-4

in the midst of great privation. It then turns to '"Merrie England" – No More!' and 'Dame Deborah Thrumpkinson, and her orphan apprentice, Joe', two short stories from Thomas Cooper's collection *Wise Saws and Modern Instances* (1845), which problematise the Victorian mantra of self-help and hard work. The chapter follows this with a discussion of 'The Charter and the Land' (1847), another anonymous short story that highlights the importance Chartism attached to an alternative, rural way of life, away from the pull of industry: as one character observes, 'Charter is the means and land the end' (Haywood 1995b, 191). Next, the chapter focuses on Ernest Jones's 'The Factory Town' (1847), a poem that highlights what was at a stake for the Chartist rank and file, whose lives were increasingly circumscribed by industrial modernity. Finally, the chapter analyses a Chartist novel, Thomas Martin Wheeler's *Sunshine and Shadow: A Tale of the Nineteenth Century* (1999, serialised 1849–50), a *Bildungsroman* and survey of Chartism's own history first published in the *Northern Star*, the movement's own newspaper (Chase 2007, 16–17).

As these examples variously illustrate, Chartist literature did not dwell on the physical realities and bodily impacts of environmental inequality; the fact of them was often assumed, even underplayed. What mattered was how the working classes responded to those conditions, the values they embraced, and the solutions they sought; what mattered was their strength and communal solidarity as a class with shared interests, a shared outlook, and shared aims. But as stories such as 'The Charter and the Land' illustrate, great importance was also attached to a future shaped by environmental goods, not harms.

*

What, then, was Chartism, and how was it related to industrial modernity? How, in turn, did the movement position itself in relation (and response) to the material circumstances (and discursive constraints) that together constituted the class-bound environmental inequalities that were already so obvious in the early Victorian period? Chartism emerged in the wake of the Reform Act of 1832, which had failed on its promise to deliver universal suffrage, and of the Poor Law Amendment Act of 1834, which established the workhouse as the only and much-hated means of providing the poor with help in times of need (Haywood 1997, 1–2). As Malcolm Chase points out, the movement has often been presented as an expression of 'the politics of hunger' (2007, 19), but it was not simply a reaction to the economic downturns and depressions that seemed to be the inevitable corollary of industrial capitalism. Hunger was an issue, plain and simple, but so too was mechanisation, which supplanted skilled, well-paid work, and drove workers into the factories, where they 'were required to submit to intrusive discipline and long working hours' (Chase 2007, 21). As Chase adds, resentment and resistance were compounded by the obvious injustice of this new arrangement; workers laboured long and hard for little if any improvement in material well-being, 'while the disparity between wealth and poverty appeared to be widening' (2007, 21). These were facts that even the well-to-do acknowledged; 'England is surely the country where luxury has reached its height and poverty its depths', wrote Florence Nightingale in 1848 (Cook 1913, Vol I, 80).

For the working classes, poverty was compounded by environmental inequality: working conditions were dangerous, living conditions polluted, unsanitary, and cramped. As one Chartist wrote of another, '[h]e lived in a cellar, nine feet by seven. *This dwelling was his workshop, his bed-room, his kitchen, his study;* AND NOT UNFREQUENTLY HIS HOSPITAL' (capitals and italics in the original; qtd in Chase 2007, 22). Thus, '[e]nvironmental degradation accentuated consciousness of the disparity between wealth and poverty'; it became 'a recurrent feature of Chartist thinking' (Chase 2007, 22). In turn, this nascent sense of environmental classism – of a pattern of environmental injustice that merged with a wider pattern of class-based inequality – was taken up by the movement's many reporters, correspondents, and writers.

As historians have stressed, Chartism was a self-consciously literary movement, well aware of the importance of aesthetic as well as political representation (Haywood 2016, 82–3; Sanders 2009, 3): from the outset, Chartism was shaped not just by mass meetings and powerful oratory but by the written word. The movement took its name from its written demands for political enfranchisement – the People's Charter was first presented at a mass rally in Glasgow in May 1838 (Chase 2007, 4, 7) – and it owed much of its sense of self to the *Northern Star*, the remarkable newspaper established by Feargus O'Connor. As Chase observes, the *Star* presented the movement as 'a coherent and vital whole … And so in time [it] become' (2007, 17). The written word was performative; like the speech acts of figures such as O'Connor, spreading the word on speaking tours (Chase 2007, 14), it enacted what it set out to describe.

As Chartists recognised, fundamental difficulties lay in what environmental justice advocates now describe as nonrecognition and misrecognition (Coolsaet and Néron 2021). Simply, the working class was ignored, or misconstrued. Lacking representation, it was itself misrepresented, as 'that great dumb toiling class which cannot speak' (Carlyle 1839 [1897], Vol 16, 207), and therefore as 'intellectually, culturally, and politically immature' (Haywood 2016, 83). The problem was, in effect, a discursive one: it relates to discourse, as the means by which those in power decide what is and can be said, and in so doing, shape what is taken to be reality, and accepted as normal. As I discussed in Chapters 1 and 2, that reality – that normative, discursive construction – was already under pressure. Much was being written about the condition of the working class; various investigations were commissioned and reports published, many of which were (in the words of a contemporary, Anna Jameson) 'teeming with graphic narratives and discoveries of horror' (qtd in Haywood 2000, 5). It was on reports such as these that Engels drew in writing his own Gothicised account of the 'Condition of England' identified by Thomas Carlyle. In turn, these reports encouraged the emergence of a new form of fiction, specifically responsive to the obvious injustice of a segregated society in which the poor endured difficult and dangerous lives in the polluted and unsanitary districts that grew up around industry. This was the 'industrial' or 'social problem' novel (Haywood 2000, 5), and notable examples of this 'reformation' of the novel (Gallagher 1985, xi) include Benjamin Disraeli's *Sybil* (1845), Elizabeth Gaskell's *Mary Barton* (1848) and *North and South* (1854–5), Charles Kingsley's *Alton Locke* (1850), and Charles Dickens' *Hard Times* (1854), a novel itself dedicated to Carlyle.

Like most (if not quite all) of the reports and investigations on which the novels drew, however, these works of fiction were written by middle-class writers, and implicitly addressed to a middle-class readership, limiting what could be and was said (Eagleton and Pierce 1979, 10–11). Certainly, these writers sought to challenge middle-class complacency, by highlighting the human and environmental impacts of the processes in which the middle classes were complicit and from which they so obviously benefited. At the same time, these writers felt compelled to reject the revolutionary implications of the situation they described, downplaying middle-class responsibility, whilst insisting that the working classes (often presented as feckless and improvident) were themselves responsible for the conditions that they endured.

Consequently, these writers tended to demonise collective attempts to improve working-class conditions, arguing that militancy (invariably presented as the work of 'a few irresponsible agitators') would 'simply retard any progress rather than advance it' (Eagleton and Pierce 1979, 16). The views of Peter Gaskell, a Manchester surgeon, are illustrative. 'It is in vain that in their rage', he wrote of the working classes in 1836, 'worked up into madness by heartless demagogues, by hunger, by the sight of their famishing children, they have taken the law into their own hands, and dreadful proof have they given how unfit they were to wield it' (*Artisans and Machinery*, qtd in Ward 1970, 21). These arguments carried over into fictional efforts to dramatise the experience of the working classes. As the Chartist writer Thomas Martin Wheeler later observed, '[o]ur novelists – even the most liberal … can never draw a democrat save in warpaint' (Wheeler 1999, 99). Eschewing direct action as a means of rescuing their working-class heroes and heroines from their polluted, poverty-stricken lives, bourgeois writers instead sought solutions in fortuitous marriages, accidents of birth, happy inheritances, or, in those situations which could not otherwise be resolved, through exile or death (Eagleton and Pierce 1979, 12). In this way, Ian Haywood observes, the conventions of the bourgeois novel (such as an emphasis on individualism or the property plot) were used to accommodate and absorb the problem industrial fiction set out to expose (2000, 6). Fundamentally, Haywood adds, these writers failed to find ways in which to identify fully with the working classes (2000, 6), perhaps sensing that the divide between them was already too deep.

At this point in its development, the dominant discourse enacted a further, crucial shift in its construction of the 'labouring poor'. In the words of Leeds surgeon Charles Turner Thackrah (1832), watching the cotton-workers leaving Manchester's mills, 'I saw, *or thought I saw*, a degenerate race – human beings stunted, enfeebled, and depraved' (italics added; qtd in Ward 1970, 29). Here as so often in middle-class discussions, the distorting and deleterious impact of industrial life on physique and health (borne out in later studies of markers such as height and life-expectancy) was confused or deliberately conflated with its impact on morals, mores, and behaviours (Chinn 1995 [2006], 79–80). Talk of debility became talk of dissipation, enabling poverty to be seen as the fault of the poor; its roots were pathological rather than economic (Chinn 1995 [2006], 94). Thus, the middle classes were absolved of involvement, just as talk of workplace 'accidents' absolved them of 'direct responsibility' for industrial injuries and deaths (Sanders 2000, 34). As Thackrah observed, Manchester mill-owners 'urged the bad habits of the Manchester poor' (qtd in Ward 1970, 29), a defence that other middle-class commentators quickly internalised. 'The real evil lies in the habits of the people themselves', wrote Peter Gaskell of the 'Manufacturing Population' (qtd in Ward 1970, 20). Even the most compassionate of middle-class writers could not entirely escape this dominant logic: in *Mary Barton*, for example, Elizabeth Gaskell attributed the 'stunted look' of factory-worker John Barton not to the systematic and systemic failures of industrial capitalism but to the accident of 'bad times' and the self-inflicted fault of 'improvident habits' (1848 [2003], 7), or as she put it elsewhere in the novel, 'child-like improvidence' (24), a twist which further positioned the working class not as moral agents but as moral patients (for a more detailed discussion of the tensions at work in the novel, see Eagleton and Pierce 1979, 34–8).

In this way, Raymond Williams observed, middle-class '[s]ympathy was transformed, not into action, but into withdrawal' (1958 [1959], 109). For Chartist activists – almost without exception working class in origin, argued Dorothy Thompson (2015, 19, 139) – it was critically important that they respond to this systematic (mis)representation, contesting its assumptions and inaccuracies, and constructing in its place a more confident, more positive account of their own experiences, their own values, and their own worth. The working classes were not, after all, simple products of their own environment; brutal as their living and working conditions no doubt were, they were not therefore brutish, and they were certainly not the 'dumb people' (Gaskell 2003, 3) of middle-class literary formulations. It is no surprise, therefore, that Chartism invested a great deal of its energy in producing literature of almost every kind, from reports and opinion pieces to poetry and forms of fiction such as the short story; exceptionally, it even took on that bulwark of bourgeois values, the novel itself. The greater surprise is that, unlike the bourgeois industrial fiction of the 1840s, these narratives have since been overlooked in mainstream literary criticism. Only a handful of critics, most notably Ian Haywood, have set about republishing these works, and highlighting their importance.[2]

My own discussion begins with 'A Simple Story' (1840), anonymously published in the *English Chartist Circular*, and reproduced in Haywood's anthology of Chartist fiction (1995b, 41–5). 'A Simple Story' immediately explains the challenge its author sets out to address: '[h]ow little of the patient endurance of misery, the real benevolence of heart, and generosity of action, of the toil-worn portion of the community, is known by the more favoured children of fortune' (Haywood 1995b, 41). As the narrator continues, precisely highlighting the problem with dominant constructions of the working classes, there is instead 'a strong propensity to associate the ideas of crime, folly and ignorance, with those of labour, manufactories, mills, and working population' (Haywood 1995b, 41). The generosity of the poor is overlooked, adds the narrator, even as the rich are lauded for 'their benevolent actions' (Haywood 1995b, 41). If only 'the kind and noble acts of the operative [might] obtain similar publicity' (Haywood 1995b, 41).

This the narrator sets out to do, by telling a short story rooted in his (or her) experience 'of those seasons of general distress and anguish' that appear all 'too common … in this highly favoured and Christian land' (Haywood 1995b, 42). In the midst of an economic downturn, when there is simply not enough work to go around, a young apprentice named Joe reluctantly admits that his whole family is going hungry; 'I did not like for any body [sic] to know how bad we were off' (Haywood 1995b, 43). As his fellow-workers belatedly discover, the family has sold off its furniture, all its clothing 'except those upon their persons', and even the mother's wedding ring; still, the family goes hungry, and still it has never 'ever been suspected of a dishonest act' (Haywood 1995b, 44). The community rallies around them, all cheerfully contributing what they can 'to the wretched family' (Haywood 1995b, 44). Yet despite its best efforts, it cannot in the end undo the damage done by these bouts of economic distress. What comes of Joe? '[W]e regret to add [the story concludes] that he died about three years after the incident occurred … from a want of sufficient nourishment, the condition of too many of the working classes' (Haywood 1995b, 45).

As this summary suggests, 'A Simple Story' offers the reader an idealised fable that deliberately contradicts the belief that the poor were improvident or immoral. Joe and his family are, the narrator insists, hard-working, self-respecting, but also caring; determined not to impose on others for support, yet always ready to offer it. They are, in the language of the day, deserving; their poverty is no fault of their own. As the story underlines, poverty is the consequence of an economic system that alternates periods of (relative) prosperity with ones of great economic distress, but always demands more of those it rewards the least. The merchants and manufacturers might face bankruptcy, but never starvation: by contrast, the narrator points out, 'millions of the working population … know by bitter experience what it is to exist months together deeply suffering from scantiness of food, fire and clothing', threatened by the very real possibility that (as the story illustrates) they might 'perish for want' (Haywood 1995b, 42).

The obvious injustice of this situation is underpinned by the environmental inequality it also implies. Well-to-do Victorian society set great store on the home, as a welcoming and supportive environment; stripped of all comforts, the home in this story becomes a cold, empty, dead space, only marginally preferable to the prison-like confines of the workhouse. There is an important, related, point, as Anna Clark points out. Domesticity (and by extension, domestic spaces) form 'an important subtext' in Chartist literature (Clark 1992, 66), since it was in the home that middle-class narratives 'demarcated the working class as different and inferior' (66), and by extension ineligible for the vote: whereas the middle-class home was the site of propriety, industry, virtue, and a clutch of respectable values that supported bourgeois rights to political hegemony, the working-class home was a site of discord, drunkenness, and self-induced misery. To the contrary, Chartist narratives insisted, the happiness of the working-class home was disrupted from without, by 'capitalism and corruption' (Clark 1992, 64). These forces transformed the working-class home not just because they introduced to it uncertainty and hunger, but because, as Clark points out, working-class 'domesticity originally meant domestic industry, when families worked together under one roof' (1992, 65). Now, industrial capitalism had itself intruded into and destroyed that ideal (and idealised) space; where it still existed, 'cottage industry had degenerated into the misery of sweat shops' (Clark 1992, 65).

Cottage industry was, in any case, in terminal decline. As Hilton points out, the factory was in the ascendant: machine-made supplanted hand-made, and factory supplanted domestic production (2006, 586). However, factory owners found it convenient to use domestic production to supplement factory production when demand was high, thereby avoiding the cost of investing in expensive machinery that might then lie idle when demand was low (Hilton 2006, 586). Consequently, domestic production 'lingered on distressingly' (Hilton 2006, 586), as Thomas Cooper (1805–92) was aware. Born in Leicester, an important centre for the hosiery trade, Cooper returned there in 1840; by then, trade was already collapsing, with dire consequences for the city's 'stockingers' (stocking knitters or weavers). Their suffering forms the backdrop to stories such as 'Dame Deborah Thrumpkinson, and her orphan apprentice, Joe' (Cooper 1845, Vol II, 150–203).

'Dame Deborah' is one of the short stories from Cooper's two-volume collection, *Wise Saws and Modern Instances* (1845), published on his release from a two-year imprisonment for the speeches he made during the 'Plug Plot' general strikes of 1842 (Gilbert 2009, 28). Cooper's story focuses on Joe, whom the Dame adopts as a child, and saves from starvation. The Dame raises Joe, establishes him in an apprenticeship, and supervises his spiritual development. Freed from indenture at the age of 16, Joe sets out to complete his education, declaring that 'man is the great book I wish to read' (Cooper

1845, Vol II, 194). Three years later, he returns to see his benefactress, Dame Deborah, buried. But 'what an altered man was Joe!' (Cooper 1845, Vol II, 202): '[a] residence in the manufacturing districts had unveiled to him a world of misery – contention – competition – avarice – oppression – and suffering – and famine – that he had never supposed to exist!' (Cooper 1845, Vol II, 202).

It is a moment of 'wrenching sadness', as Gregory Vargo points out (2011, 172), reinforced by the double movement it depicts: in true Victorian style, Joe has pursued his own path to 'independence and self-sufficiency', but in doing so, he has severed his links with the community that sustained him (Vargo 2011, 169). Moreover, his pursuit of understanding has brought him only the disillusioned realisation that, writ large, this same search for autonomy – for an illusory 'vista of possibility' (Vargo 2011, 169) – has brought about a fractured, atomised society, in which the shibboleth of 'self-help' simply serves to mask self-interest. For Joe, therefore, 'individuation is not self-realisation, but loss', a loss not only of community but also of hope (Vargo 2011, 169).

Joe's trajectory highlights two, important points. The first is that individual agency is not and never will be sufficient to mend a stricken society – a society itself undone by fictions of self-sufficiency. Hope rests in social agency, and 'mass politics' of the kind that Cooper himself sought to sustain in his native Leicester (Vargo 2011, 169). The second point relates to the elision that Cooper's story also enacts. Joe's story is carefully told, but at the very moment his adult life begins, it stops, abruptly; Joe's experiences in the nation's new 'manufacturing districts' are reduced to a shortlist of suffering ('avarice – oppression – and suffering – and famine') that leaves no room for explanation or description. Perhaps Cooper felt that there was no need for it, knowing that his readers would understand all too well what was meant by the reference to the 'world of misery'. Perhaps Cooper preferred to downplay those conditions, since they were not the sum of the working-class communities that struggled with them. Or it may be that, as Haywood suggests in relation to the next story under consideration, it relates to the inherent limits of nineteenth-century realism as the embodiment of 'a bourgeois view of the world, with property and marriage the driving force of most storylines' (1997, 7).

'"Merrie England" – No More!' (Cooper 1845, Vol I, 201–17) is another of the stories that make up Cooper's collection. 'Relentlessly dialogic' rather than descriptive (Vargo 2011, 175), it centres on a group of out-of-work and impoverished Leicester stockingers, who restlessly debate their predicament and how best to address it. Religion seems to preach quiescence, and the hope of a better life to come (Cooper 1845, Vol I, 203–4; 213–4), but in the here and now, hunger is the pressing problem; if moral force cannot persuade society of the need for a meaningful solution to it, then perhaps physical force will be necessary (Cooper 1845, Vol I, 209). Matters come to a head when the men confront the 'red-coat' (Cooper 1845, Vol I, 207) (or soldier) who is about to recruit one of their sons ('a pale, tall stripling, who seemed to be but sixteen or seventeen years of age' (206). To the crowd that assembles around the soldier and his prize, soldiery is a still worse form of slavery than the one they daily endure. There is a brief confrontation; the soldier thinks better of it, and retreats; and the stockingers contemplate their moment of triumph. 'Our eyes are getting opened', declares one; 'they may be able to kill us off by starvation, at home; but I hope young and old will have too much sense, in future, to give and sell their bodies to be shot at, for tyrants' (Cooper 1845, Vol I, 208).

Yet starvation remains an overriding concern. The story concludes with a scene reminiscent of the one in 'A Simple Story', as the stockingers gather in mutual support. A young woman is on the brink of starvation: she can no longer feed her baby, her husband can find no work, and the parish will provide him with no relief;

> to complete their misery, they owed six weeks' rent for the room in which lay the bag of shavings that formed their bed; and, if they could not pay the next week's rent, they must turn out into the street, or go into the Bastille.
>
> (Cooper 1845, Vol I, 212)

The scene is played out in a 'small wretchedly furnished habitation', itself a mocking echo of the bourgeois emphasis on home and hearth, and so 'squalid-looking' (Cooper 1845, Vol I, 211) that the reader is left to wonder at conditions in the 'Union Bastille' (or workhouse), which themselves invite such widespread horror. In the meantime, their 'oppressors ride in their gigs, and build their great warehouses' (Cooper 1845, Vol I, 213). The stockingers do and give what they can. Here, the narrator confronts the reader directly, and challenges her or him to doubt the truth of what he is presenting.

> These conversations are *real*: they are no coinages. Go to Leicester, or any other of the suffering towns of depressed manufacture, where men compete with each other in machinery till human hands are of little use, and rival each other in wicked zeal to reduce man to the merest minimum of subsistence.
>
> (Cooper 1845, Vol I, 215)

This misery cannot be mended, the narrator insists, without real, and radical change; there will certainly be no answers for the suffering people he depicts, who will go on 'starving, – begging, – receiving threats of imprisonment'; finding 'a little work', and then starving again, 'still going the same miserable round' (Cooper 1845, Vol I, 217).

'What are your thoughts, reader?' (Cooper 1845, Vol I, 217). With that question, the narrator abruptly ends the story. It is, Haywood argues, a striking conclusion to a highly unusual narrative (Haywood 1997, 6–7; 1995a,

8–9); as Haywood adds, the story constitutes a kind of 'anti-realism' (2016, 96) set up in deliberate opposition to realist, bourgeois forms of fiction, with their emphasis on the 'almost obligatory' happy ending (Eagleton and Pierce 1979, 12). In 'Merrie England', Haywood argues, '[r]esistance to narrative closure is … the only "realistic" option' (1997, 7), and as '[t]he prose collapses into a string of disconnected fragments, a miserable "round" of mere sequential events', it becomes clear that realism 'has reached the limits of representation' (1995a, 9).

Haywood's observation is intriguing, and it gives a further insight into the nature of the story, in which very little happens, but much is said. As Haywood points out, 'the point of [Cooper's] approach to his material is to demonstrate the intelligence of these self-educated workers' (1997, 6): 'whoever enters Leicester', declares the narrator, 'or any other of the populous starving hives of England, must expect to find the deepest subjects of theology, of government, and political economy, taken up with a subtlety that would often puzzle a graduate of Oxford or Cambridge' (Cooper 1845, Vol I, 216). Thus, Cooper (himself a prodigious autodidact [Sanders 2009, 7–8]) sets out to contradict those who would decry the working classes as an ignorant mass, ineligible for the vote. At the level of discourse, therefore, 'Merrie England' counts as another of those working-class interventions in a debate then dominated by middle-class constructions of the poor as the deprived, but also potentially corrupt products of their environs; Cooper insists that, to the contrary, his characters (thinking, feeling, animate individuals) bear little or no resemblance to these clichéd formulations, with their roots in the logic of environmental determinism.

However, Cooper never claims that his characters have somehow transcended their own bodily realities, their own lived experience; they are cruelly bound up in their own corporeality, trapped (as here) by their own hunger, or, as Cooper simply remarks in another story about Leicester's workers, 'suffering from want of food, fuel, and clothing' ('The Minister of Mercy', Cooper 1845, Vol I, 197). The point, his narrative implies, is not that the material world is irrelevant; it all too obviously influences the discursive domain that these workers contest as they wander, hungrily, through the streets, in search of work or relief; the point is, perhaps, that the realities of their lived experience eludes realism itself. It does not bear representation; it eludes portrayal.[3] If what is in prospect is the horror of death itself, this is perhaps an obvious statement, but the reality to which Cooper gestures – the unfolding of 'slow violence', in Rob Nixon's phrase (2011) – amount to a kind of death-in-life: a hand-to-mouth existence in which work alternates with unemployment or short-time labour, and hunger haunts even those times of relative plenty. How can such a protracted, desperate state be represented, except by allusion, or by obvious omission, or perhaps even, as here, by contrast with its alternate? The story's title is, after all, '"Merrie England" – No More!', at once a sarcastic comment on the state of a supposedly great and prosperous nation at the forefront of humankind's headlong fall towards industrial modernity, and a glance backwards to the kind of pastoral illusion whose appeal (as I discussed in the previous chapter) even Engels cannot escape.

The question of whether the realist literary mode – and by extension the realist novel – has the capacity to represent these lived experiences is one to which I will return later in the chapter. The next of the stories on which I focus takes up the myth of a free and prosperous 'Merrie England', and explores the way in which Chartism itself sought to turn it into a reality. 'Nature was exhausted!' exclaims the narrator in '"Merrie England" – No More!' (Cooper 1845, Vol I, 211), referring to the young woman who cannot even feed her baby, but it is not just human nature that industrial modernity is expending. What, then, would the world look like if political economy was dismantled and industrial modernity arrested or reversed? 'The Charter and the Land' takes up this theme. As Haywood points out, 'The Charter and the Land' was written purposely 'to promote the National Land Company' (1995a, 20), which aimed to provide working-class people with enough land to farm and become self-sufficient: it would, in other words, provide the working classes with an alternative to factory life, whilst also enfranchising them by virtue of the property they would be awarded (Chase 2007, 254). As Andrew Messner points out, the Land Plan was therefore central to the movement's concept of an alternative political economy, and it was all the more important and innovative because it sought to combine 'the radical ideal of independence' with a 'culture of mutuality', making it 'one of the most significant co-operative experiments of the nineteenth century' (1999, 1096).

At one level, notes Rob Breton, 'The Charter and the Land' is 'unabashed propaganda' (2013, 139), but at another, it dramatises something of the human cost of activism, just as it also underlines the need for a future on to which Chartists might project their hopes. 'The Charter and the Land' tells the story of William and Betsy Wright and their 'two little factory children' (Haywood 1995b, 191), a family of working-class weavers (192) in industrialised Stockport. William has taken up the cause of Chartism, and as was the case with so many of the movement's members – rank and file along with leadership – he is imprisoned 'for what is termed "sedition, riots, routs, and tumults"' (Haywood 1995b, 191). Ruined, without character, and therefore unemployable, he cannot support his family; the task falls to his wife, Betsey, whose only concern when she next gives birth is 'how soon she might leave the baby and return to the Mill' (Haywood 1995b, 191); middle-class readers may well have found these details particularly disconcerting, given the importance they attached to the sacred role of wives and mothers as 'Angels in the House' (Langland 2001, 119). Confronted by this grim reality, it seems that William has 'somewhat predictably' taken to the bottle (Breton 2013, 139), abrogating his responsibilities as husband and father. In fact, he is scraping together just enough money to invest in the 'Chartist Co-operative Land Company' (Haywood 1995b, 191), or National Land Company, as it would become in 1846 (Breton 2013, 139).

It will transpire, of course, that William is one of the lucky few to be awarded land. Betsey is not immediately convinced. She wonders whether working the land will be any better than factory life, or any more financially stable (Haywood 1995b, 192). This was a real concern amongst Chartists; as this exchange between Betsey and William illustrates, the story sets out to provide reassurance.

> 'Why, thour't not used to land, and folk say it'll harass and kill thee'. 'Why', replied Will, 'father, and grandfather, and all folk belonging to I worked on land and it didn't kill them, and why should it kill me? I worked too when I was a lad, and I was stronger and heartier than I am now ...'
>
> (Haywood 1995b, 192)

As many working-class readers would have agreed, this was not an unreasonable response. Although a narrow majority of Britons were now living in towns and cities, as the Census of 1851 established, most were also migrants from the country. Many would indeed have been familiar with working the land – and aware of its hardships. As B. W. Clapp points out, sanitation was no better in the country, housing sometimes worse, and diet poorer (1994, 65–6), and on that basis alone, many continued to migrate to the city (65). But as they then discovered, life in the factories and urban slums introduced new and unwelcome stresses that were at once environmental, economic, psychological, and nutritional (Luckin 2015, 173), and it was the complex combination of these stressors that explains the otherwise confusing disparity in life expectancies between city and country: as Romola Davenport points out, 'death rates were generally higher in urban areas compared with rural ones, a phenomenon dubbed the "urban penalty"' (2020, 455). For example, '[c]hildren born in Manchester suffered a penalty of 16.4 years compared to the average for England as a whole' (Voth 2004, 285), but the differential was sometimes still higher when translated into an urban-rural divide: in 1851, for example, life-expectancy in Liverpool – a centre of maritime trade as well as industry – was a mere 26, but in rural Okehampton, it was 57 (Daunton 2000, 63).

When in 'The Charter and the Land' William Wright claims that life in the country is hard-working but healthy, he is, therefore, arguing from experience, and speaking what was at least a partial truth. Moreover, he is not suggesting that, in returning to the land, factory workers leave one master for another; the Land Company provided the fortunate with land and a house, in communities they had a share in running. Like 'emancipated slaves', the Wright family settle in their 'Holy Land' (Haywood 1995b, 193), there to take up a life that is healthy and, as we would now recognise, sustainable; (environmental) justice has been done, both at the local level, and at the larger, parliamentary level, in which William would (as property owner) also have had a say. Thus, the Chartist attempt to found 'a Small Proprietary Class' (Haywood 1995b, 193) had implications that were politically and economically sweeping, giving workers the vote whilst reducing the pool of 'debased and brutalised' labour (Haywood 2000, 7) on which industrialists might draw. As William points out, 'I'll be one less for tyrants to fall back on to reduce wages' (Haywood 1995b, 192). Neither he, his wife, nor his children will have to return to the factory; as his young ones joyously declare, swopping factory floor for lessons in the community's school, 'Mammy, sure you won't let us go back to Stockport and factory any more to be whipped' (Haywood 1995b, 193). (In a moment calculated to touch the story's readers, Wright's family mistake the school bell for the factory bell, and in so doing, confuse the possibility of a better, and better-educated future, with their tyrannised and brutalised past [Haywood 1995b, 193]).

The Land Plan was not, therefore, an escape from the flow of history, but an attempt to advance it, by pursuing a 'natural' wage independent of the traditional hierarchy of landed rule (Vanden Bossche 2014, 123), and a form of unalienated labour beyond the reach of the marketplace or the machinations of industry: 'God bless LAND and the CHARTER' (capitals in the original; Haywood 1995b, 194). In practice, however, the plan was beset by challenges from both within and without. Many did not take to life on the land, and in any case, there was too little of it to make a real living possible (Chase 2007, 353–4). Moreover, the government was intent on undermining the legitimacy of the Land Plan, and pointed out that it was, in effect, a lottery – families invested, lots were drawn, and only a few were lucky – and consequently illegal. Yet there were those amongst the fortunate who, like the writer Thomas Martin Wheeler, took to a life of spade husbandry, and recognised its utopian potential (Wheeler 1999, 176–8), as a direct expression of environmental justice, enacted and embodied in a practical, communal way. These arguments were not forgotten later in the century, as Dennis Hardy points out; some 20 years later, 'the banner of the land' was taken up by noted thinker John Ruskin (Hardy 1979, 78), and whilst Ruskin's attempts to forge feudal communities were largely unsuccessful, a more popular and egalitarian 'back-to-the-land' movement also emerged towards the end of the century (81); as Peter Gould argues (1988), this movement itself constituted an early form of green political activism.

The power of this vision of a cooperative, agrarian community re-surfaces in the work of Ernest Jones (1819–69), one of the last of the Chartist leaders, and Cooper's successor as 'the movement's leading poet' (Sanders 2009, 80). It has been argued that poetry as a form was ill-suited to social engagement (see Sanders 2009, 29–30), but the Chartists used it as a potent means of articulating their sense of injustice, sometimes on a grand scale; Cooper himself contrived to write a ten-book poetic epic entitled *The Purgatory of Suicides* (which included a sonnet dedicated to Carlyle) whilst in Stafford jail (1843–5; Gilbert 2009, 27–8). My own focus is on Jones's 'The Factory Town', collected in *The Battle-Day, and Other Poems* in 1855 (Jones 1855, 82–8; see also Ward 1970, 32). Perhaps more graphically and insistently than any of the texts I have discussed so far in this chapter, Jones's poem emphasises the nature of environmental injustice as it emerged in the early Victorian period. In doing so, however, Jones's poem also highlights two other aspects of the working-class experience. One is a subtler sense of the risks to which the working classes were exposed, risks that were themselves elusive, as Ulrich Beck predicted (1992 [2012], 22–3). The other is a strong and shared belief that solutions could only lie in a radical break with industrial capitalism, and the structures of meaning encoded in modernising discourses: in this sense, Chartism's opponents were right to recognise its revolutionary intent (Foot 2005 [2006], 118–9). At the same time, however, Jones's verse enacts a further, decisive shift that disconnects it with Chartist counter-narratives: it reprises, even if only in passing, the deterministic narrative of demoralisation and degeneration, the very trap that other Chartists writers had sought to undermine and whose premise they set out to counter.

Over the course of its 27 stanzas, Jones's poem insists on the appalling realities of life in the factory town: of the grim realities of the workplace, and the worse nature of slum life. For example, the poem repeatedly highlights the unrelenting tempo of factory work, driven by machines that need never stop: 'factories gave forth lurid fires', the narrative explains, but '*man's* volcanoes never rest' (italics in the original; Jones 1855, 82). Elsewhere, the poem refers to 'shrunken sinews quivering / To the engine's horrid beat' (84). All are now subject to the machine:

> Women, children, men were toiling,
> Locked in dungeons close and black,
> Life's fast-failing thread uncoiling
> Round the wheel, the *modern rack*!
>
> (Italics in the original; Jones 1855, 82)

In this Dantean inferno of 'burning wrath' (Jones 1855, 82), industrial modernity has created new instruments of torture. Yet the factory owners are not inclined to care: masters contemplate 'factory scenes' and see in engines and (now reified) workers 'nothing but *machines*' (italics in the original; Jones 1855, 84).

For these workers, there is no meaningful escape from the factory bell that so torments the Wrights. They are entrapped within the factory town, a confinement whose impact is both psychological and physical; as Jones's poem memorably dramatises, the workers (men, women, children) yearn not only for basic necessities, but for more intangible (and no less essential) environmental goods, the want of 'air, and light' (Jones 1855, 87):

> Oh! for but one drop of water!
> Oh! for but one breath of air!
> One fresh touch of dewy grasses,
> Just to cool this shrivelled hand!
>
> (Jones 1855, 84–5)

In highlighting these realities, Jones also tries to reference the nature of the threat that the factory town embodied. The machine-driven rhythms of working life are one aspect of this risk to a flourishing human existence, but the poem also carries a sense of other, more elusive threats, for which it is difficult to find the right term or phrase, perhaps because these threats are themselves not properly or fully understood; indeed, they exist at or beyond the limits of scientific (and human) comprehension (Beck 1995 [2002], 9). For example, the poem refers to the city as a bubbling cauldron, boiling over with poison, where 'reeking walls environ / Mingled groups of death and life' (Jones 1855, 83). Elsewhere, it twice invokes 'fever' (83, 88) as a marker of these new risks (as I discuss in more detail in Chapter 4, 'fever' stood for a whole range of poorly understood diseases that were the inevitable counterpart of life in the crowded and unsanitary quarters of factory towns). But the poem also references smoke and fumes (Jones 1855, 83), which were themselves prominent environmental hazards in a factory town. The most famous of smoke-bound cities was, of course, London, but in this new industrial age, smoky skies were an increasingly common and widespread problem. Manchester was '[l]ong notorious as the centre of a large area blackened by smoke' (Clapp 1994, 48); in the novel *North and South*, Elizabeth Gaskell placed it at the centre of an appropriately named 'Darkshire' (1854–5 [2000], 38). Around Dudley, noted the engineer James Nasmyth during a visit in 1830, the smoke of furnaces 'blackened the country as far as the eye could reach' (1883, 167). Inevitably, this lack of clean, fresh air had a pronounced impact on health, generating respiratory problems and shortening lives in ways that were obvious even to the early Victorians (Clapp 1994, 65): '[f]or most of Victoria's reign', Clapp argues, 'lung diseases caused a quarter of all deaths' (67).

Caught, therefore, between poisons, fever, and fumes, and trapped within the factory towns that were the concomitant of industrial modernity's dominance, the working class no doubt felt (as Jones puts it) like the cursed 'sons of *Cain*' (italics in the original; Jones 1855, 85), an unsettling Biblical reference that invokes Cain's exile in Eden's barren antithesis, the Land of Nod. As Jones wrote, 'a banished population / Festers in the fetid street' (1855, 88). Where might hope lie? Jones repeatedly contrasts factory life with rural life on the land, even as he highlights the encroachment of one on the other:

> Thinner wanes the rural village,
> Smokier lies the fallow plain –
> Shrinks the cornfields' pleasant tillage,
> Fades the orchard's rich domain;
>
> (Jones 1855, 88)

As the poem continues, in a pointed reference to the role of cotton manufacture as an economic driver, 'Give us, God, to save our nation / Less of *cotton*, more of *wheat*' (italics in the original; Jones 1855, 88).

There is, however, more to this contrast than a recapitulation of the pastoral mode, with its contrast between town and country: as the poem then adds, 'Take us back to lea and wild wood / Back to nature and to Thee!' (88). Beyond the pastoral, with its suggestion of 'pleasant tillage' and 'rich domain', the poem invokes 'wild winds' (Jones 1855, 82, 88), 'shady forest land' (85), and the enchantments of stars, moon, and 'fairies' mummer' (85); it co-opts, in other words, the conventional sense of a Romanticised 'Nature' – 'immutable and good' (Beck 1995 [2002], 39) – as a counterpoint to life in a factory town. This 'Nature' is the one with which well-to-do Victorians were familiar: it is the idealised, poeticised 'Nature' of the Lakeland poets, Wordsworth (then Poet Laureate) pre-eminent amongst them. Why does Jones invoke it? For working-class readers, it might be argued that this 'Nature' had much less meaning (and value) than practical appeals to the land. Were these allusions, perhaps, an unwitting reflection of Jones's own unusually affluent origins? Jones was, in fact, a well-connected barrister, but as Thompson acknowledges (2015, 140), he was nothing if not committed to the cause: imprisoned as an activist, he was treated more harshly because 'seen as a class-traitor' (Gilbert 2009, 28).

The point, as Chase adds, is that Jones consciously spoke for and to a movement that saw itself as bigger in aim and outlook than sectional, class-bound politics (2007, 359), and as one of the movement's last leaders, he would later advocate the need to draw on middle-class support (355). Furthermore, and whilst these references to the Romanticised 'Nature' of lake and mountain are undeniably conventional, they reflect an important truth that studies have variously reaffirmed in the last few decades: that encounters with the kind of wild or wilderness nature to which Jones refers is itself an important (if only ever temporary) antidote to the alienations of modern, urbanised existence (for a critical engagement with this idea, see Walton 2021). Indeed, Martha Nussbaum lists living 'with concern for and in relation to animals, plants, and the world of nature' (2006, 77) as one of the core human capabilities that society must foster as 'part of a minimum account of social justice' (75). As such, these moments of contact are themselves a constituent part of a flourishing existence, to which every individual has a right (Miles 2022, 18).[4] Moreover, and in the context of a sense of environmental justice that extends beyond narrowly utilitarian needs and anthropocentric formulations, these moments presuppose the existence of nonhuman, natural processes beyond reach of both the factory towns and the cultivated, constructed world of field and farm.

There is, nevertheless, a further and more problematic shift in Jones's poetic reasoning, captured as he insists once again that those immured by factory life are:

> Perishing for want of *Nature*!
> Crowded in the stifling town –

> Dwarfed in brain and shrunk in stature –
> Generations growing *down*!
>
> (Italics in the original; Jones 1855, 87)

In these lines, Jones acknowledges the bodily impact of industrial modernity, and the no less real impact of the abnatural (Taylor 2016, 5), the term that Jesse Oak Taylor uses to denote nature's erasure, absence or withdrawal (for further discussion of this term, see Chapter 2 of this book). Jones also notes the possibility that these effects might be carried over into future generations. In making his point, however, Jones takes another step, and with the reference to '[d]warfed in brain and shrunk in stature', opens the way to deterministic narratives about demoralisation and degeneration. It is the trap which, as we have seen thus far, other Chartist narratives ignored or strenuously sought to subvert: that industrial modernity was in effect creating a people which was not just physically less robust or healthy, but somehow lesser, a people whose plight the middle classes might acknowledge, but whose right to political participation and social involvement could be dismissed on the same grounds – that the working classes were in some sense (morally and intellectually as well as physically) becoming inferior and 'other', and therefore undeserving of consideration as full and free members of society.

Part of Jones's difficulty, it might be argued, lies in the high-flown diction and lofty tone that he necessarily adopts as part of the conventional poetic mode, a language of 'high-cultural Romanticism' (Ledger 2002, 32) that working-class readers might themselves have found alienating (33). For all the poem's earnest indignation, it condescends to and generalises about its subjects, whose own complex and variegated experience is subsumed in generalisations about 'the people' as homogenous mass. As Val Plumwood has argued, the process of homogenisation is itself a first and steps towards a dualistic 'othering' or marginalisation (1993 [1997], 53–5), and in this respect, even the briefest of Chartist short stories may have been a more effective medium for illustrating the rich and complex experience of early Victorian workers. This is not to dismiss poetry *tout court*, or discount Chartist poetry in its entirety; in *The Poetry of Chartism: Aesthetics, Politics, History* (2009), Mike Sanders makes a compelling case for the effectiveness of Chartist verse. However, Jones's poem does highlight the potential pitfall of a form so closely allied with bourgeois values, and more broadly, of negotiating with any form of literary expression when those values have become so dominant a part of society's literary self-expression.

What was true of poetry might also be said of the nineteenth-century novel. As I noted earlier, the realist novel was the pre-eminent form of bourgeois literary expression, closely associated with the middle-class displacement of aristocratic patronage as a cultural determinant (Morris 2003 [2011], 77–8); in the form of the industrial novel, it offered an important if flawed commentary on the 'Condition of England'. Given those limitations, it is perhaps not surprising that would-be Chartist writers chose other forms of expression. It is certainly true that few Chartist novels were written during the period, although the more likely reason is that few working-class activists had the time for so substantial an enterprise. Yet some did, perhaps recognising, as Sally Ledger points out, that poetry was losing its cultural primacy to fiction (2002, 36; see also Sanders 2009, 82–4). Cooper himself claimed to have written – and then lost – the manuscript for a novel (Haywood 1997, 7); it has never been found. However, 'late' Chartism did generate several other, full-length novels (Haywood 2016, 93), two of which Ian Haywood has reprinted (1999; reprinted by Routledge, 2016). In the final section of this chapter, I discuss one of those novels, Wheeler's *Sunshine and Shadow*, which was itself a deliberate challenge to 'the bourgeois hegemony of literature' (Haywood 2016, 85), and a conscious attempt to wrest 'supremacy in the novel' (as Wheeler himself put it in the first instalment) from 'the opponents of our principles' (1999, 72).

How, then, does Wheeler's novel find its own form, and in so doing counter the deterministic generalisations of the day? As Ian Watt observed, the realist novel is characterised 'by the amount of attention it habitually accords both to the individualisation of its characters and to the detailed presentation of their environment' (1957 [2000], 17–18), features that would in theory enable the novel to depict environmental inequality as it registered in the actual, lived experience of working-class communities. Indeed, this characteristic tendency towards particularity and individuation – or, in Derek Attridge's phrase (2004), singularity – might itself act as a corrective to the 'logic of colonisation' (Plumwood 1993 [1997], 43), with its construction of the homogenous 'other' as 'that which the existing cultural order has to occlude in order to maintain its capacities and configurations, its value-systems and hierarchies of importance; that which it cannot afford to acknowledge if it is to continue without change' (Attridge 2004, 30). As Attridge adds,

> [t]o respond fully to the singular otherness of the other person (and thus render that otherness apprehensible) is creatively to refashion the existing norms whereby we understand persons as a category, and in that refashioning – necessarily inaugural and singular – to find a way of responding to his or her singularity.
>
> (2004, 33)

Thus, literature is itself a singular, and singularly inventive response to alterity, a creative refashioning or performance; and it is this that Wheeler's novel sets out to enact at the level of both form and content.

Sunshine and Shadow is a hybrid, combining a *Bildungsroman* (a staple of the Victorian novel) with several other genres. Ian Haywood identifies four: 'the serialised novel; the "improving", respectable, didactic Evangelical tale; the sensational, melodramatic, Gothic "penny" dreadful, and the radical pamphlet and journal' (1999, x). One might also add both the romance (since there are moments in the novel that depend on the uncommon and exceptional) and the romance novel to the list, since the romantic element within the narrative is a commanding part of it. This heady mix provides Wheeler with the tools to tell a gripping, powerful tale, but it also underpins a polemical, populist mode of address, 'consonant', as Ledger has argued, 'with the vocabulary of class identity and class conflict' that emerged 'as a powerful discursive field' within later Chartism (2002, 32).

These elements are focalised through the story of the novel's protagonist, Arthur Morton, a story that over the course of 37 instalments highlights both an obvious truth and its corollary: that if the working classes cannot be reduced to the sum of their conditions (no matter how filthy or polluted) or their predicament (no matter how hungry or desperate), nor are they simply the sum of their democratic struggles (whether those make them heroes or rebels). There is much more to a Chartist than the Charter. In his novel, Wheeler tries to give his readers a comprehensive account – a complete reading – of his flawed hero's life, in a fully realised narrative that charts Arthur's progress through childhood into maturity. Arthur's life is nothing if not eventful, taking in shipwreck, exile, imprisonment, a doomed love-affair, and of course, Chartism itself, with which his life intersects at various key moments in the movement's history. (Thus, Wheeler manages to give the reader a history of the movement as well as a 'history' of his protagonist.) In *Sunshine and Shadow*, however, the crux of Wheeler's message lies not in the exceptional but in the ordinary circumstances of working-class life, for example, when Arthur is able to marry Mary and finally settle (Wheeler 1999, 156–7). It is, as the novel's title suggests, a life of sunshine and shadow, 'now gloriously bright, anon fearfully dark', a pattern repeated time and time again in the experience of 'the sons and daughters of toil' (Wheeler 1999, 158). It is in their 'strange truths', Wheeler insists, that the reader may find 'a truer picture of humanity … than can ever be gleamed from the thousand fashionable novels' of the day (1999, 158).

The 'truer picture of humanity' embodied in Arthur's life story enables Wheeler to emphasise the point, noted earlier, that political enfranchisement is only the first step towards a solution that must also extend to radical, socioeconomic reform; *Sunshine and Shadow* contains its own panegyric to the virtues of the Land Company, land ownership, and the benefits of the smallholding as antidote to industrial modernity (Wheeler 1999, pp. 182–3). The novel form also enables Wheeler to develop another argument, which relates – but also runs counter – to Chartist portrayals of an invariably virtuous working class, transcending its own bodily realities through mutual assistance and care. Hunger may drive any man to crime, Wheeler argues, but the real crime is that a prosperous society allows hunger to exist (1999, 162). Briefly, Arthur and Mary enjoy a settled, stable, and prosperous life, supported by their own industry. Then, economic depression deprives them both of work (Wheeler 1999, 160–1). Selling everything they possess, they abandon their cottage for a succession of steadily more squalid domiciles, 'each one more dreary and comfortless than the last' (Wheeler 1999, 160). Finally, there is nothing left to them but the 'courts and alleys' of London's 'rookeries' or slums (Wheeler 1999, 165), an 'abyss of dark and unutterable suffering' (162). The only alternative is the workhouse, but to be 'environed by such indignities' (Wheeler 1999, 162), is no alternative at all; those who take this step, Wheeler insists, 'are for ever branded' (163) as 'the offal of humankind' (166); 'we speak from experience, and know that it is true' (163).

Arthur is now at his lowest ebb. Close to starvation, and in despair at his inability to support his family, Arthur is driven to 'crime and drunkenness' (Wheeler 1999, 166). Yet Arthur is not 'naturally depraved' (Wheeler 1999, 166), any more than the working class is naturally predisposed to crime (166). To the contrary, the evidence of his life story (as that of so many other working-class narratives) suggests that he is an upright, respectable man, who has, exceptionally, been driven by adversity to do desperate things. But hard times are no accident, as the narrator insists. They have a cause; and 'we can now trace the evil to its true source – the unequal distribution of wealth' (Wheeler 1999, 166). To die of hunger in a wealthy society, as thousands do every year, is to be murdered by it; it is to be 'murdered by the vile ordinances of society' (Wheeler 1999, 162). It is a revolutionary conclusion, but amongst Chartists, it was a widely shared opinion. As Engels had himself observed, '[t]he English working-men call this ["death from starvation"] "social murder", and accuse our whole society of perpetrating this crime perpetually. Are they wrong?' (Engels 1993 [2009], 38).

Encoded within the narrative, there is a further, notable premise, to which Josephine McDonagh draws attention in her analysis of 'unsettlement' (2021, 219) in Victorian society. The London to which Wheeler draws attention has become 'a hungry monster' consuming 'the bodies of hapless migrants' (McDonagh 2021, 1), always and invariably denying them the opportunity to settle or, in the terms of ecocritical discourse, to dwell, meaningfully and mindfully entangled in environmental relationships of care and consideration (arguably, it is only through just such a mindful relationship that a sustainable existence can ever be achieved; see Garrard 2012, 117). In Wheeler's novel, the impact of this pattern of unsettlement is underlined by his careful description of the suburban oasis that, briefly, Arthur and Mary establish (Wheeler 1999, 156–8), only for their contented little world to be unpicked piece by piece through poverty. Things matter a great deal to Arthur, hitherto 'thrown a wanderer on the wide world' (Wheeler 1999, 157); a cottage 'neatly but plainly furnished' (156) becomes the space within which they can develop their shared interests (in music, literature), debate their democratic ideals – and, in due course, raise a family. As McDonagh insists, it is a space they inhabit rather than occupy or appropriate (2021, 185), a space in which a carefully tended garden (Wheeler 1999, 157) plays a modest but important role, as partial antidote to the smoky and noisy built environment into which Arthur every day plunges for work.

There is a related point. For Arthur, the injustice of his situation is made manifest by the contrast between this 'peaceful abode' (Wheeler 1999, 158), the 'wretched and filthy' (165) conditions he and his family must now endure, and the 'snug domiciles' of the 'rich and respectable', on whom the depression has seemingly taken no toll (165). 'How bitterly the extreme of want is felt when surrounded by opulence and plenty', the narrator observes, 'how hideously grand seemed the splendid domain of Belgravia to the hunger-pained artisan as he passed its splendid mansions' (Wheeler 1999, 161). This obvious injustice stands for the wider problem, of a class that is marginalised, ignored, and ceaselessly exploited, and of a problem which enfranchisement will not in itself solve. This is Wheeler's second, no less radical conclusion, and in the novel, it is highlighted when Arthur settles in America for 18 months (Wheeler 1999, 139). Here, there is political liberty, but as Arthur discovers, 'political liberty was only valuable as a means to an end, that in itself it was powerless against the spirit of competition' (Wheeler 1999, 139). 'There, as in Britain, the mass of the population was at the bottom of the wheel – the many dependent upon the capital of the few' (Wheeler 1999, 139). Political enfranchisement is not in itself an antidote to socioeconomic inequality, Arthur realises, nor a solution to the inequities created by unconstrained industrial capitalism; '[t]he aristocracy of wealth was becoming as dangerous as the aristocracy of rank' (Wheeler 1999, 139).

As contemporaries of every political persuasion recognised, it followed that, for the Chartists, political emancipation was only the first step to a much more radical reform of the basis of society. To be a Chartist was 'to court persecution, transportation, and death' (Wheeler 1999, 92) because, as Paul Foot has argued, 'the Charter meant the granting of much more than a right to vote – far too much at any rate for the rulers even to contemplate' (2005 [2006], 119). 'Therein lies the difference between Chartist democracy and all previous political bourgeois democracy', wrote Engels (1993 [2009], 242):

> Chartism is of an essentially social nature, a class movement. The 'Six Points' which for the Radical bourgeois are the beginning and end of the matter, which are meant, at the utmost, to call forth certain further reforms of the Constitution, are for the proletarian a mere means to further ends. 'Political power our means, social happiness our end', is now the clearly formulated war-cry of the Chartists … they must either succumb to the power of competition once more and restore the old state of things, or they must themselves entirely overcome competition and abolish it.
>
> (Engels 1993 [2009], 242–3)

Notably, Wheeler's response to the American experience also extends to a personal vision of what a just society might look like – a vision, the narrator claims, that 'will throw into shade ... all the visions of the Utopians' (Wheeler 1999, 140). Reflecting the essentially moral argument that Chartism made, it is a vision of a society shaped by 'impartial law' and 'paternal state' (Wheeler 1999, 140), which allows for individual liberty whilst also fostering a flourishing life for all; its aim is to ensure and enable 'the free development of [each individual's] improvement and happiness; each one respecting himself and his fellows as moral beings' (Wheeler 1999, 140). In a gesture towards a more expansive sense of justice that extends moral consideration beyond species boundaries, Wheeler also insists that there must be '[n]o oppression bowing down the weak to the strong – the friendless to the favoured; no lordship of man over any of his species; no woman bowing beneath slavery's yoke; no power but for the common weal' (Wheeler 1999, 140). Society must be governed, that is, by the recognition that the 'common weal' depends on a wider sense of flourishing, which takes in nonhuman as well as human agencies, without 'lordship' over any species. It requires 'verdant oases' beyond reach of both the factory and what Wheeler described as the 'agricultural desert', a phrase that tellingly anticipates the reality of today's farmed monocultures (1999, 182); it requires not the 'universal blank' of large fields lacking any form of biodiversity, but 'the delight of hill and valley, wood and stream, tower and waterfall', and all the melodies that accompany them (183). It requires, in other words, the kind of constitutive contact to which Ernest Jones also drew attention, which in turn necessitates an emergent form of justice that escapes the gravitational pull of human affairs, with its 'chaos and confusion', its cruelties and injustices. A truly just society is one that imitates those 'natural laws' that foster growth – in the natural world, Wheeler elsewhere writes in the novel, 'every tree, every plant, is placed in a soil and situation suited to the development of its peculiar properties' (1999, 170) – and an untouched, self-determining natural world is itself an integral part of what constitutes environmental justice. Yet that natural, nonhuman world was itself under threat. '[I]n a country like Britain', Wheeler wrote with great perception, 'where capital and its interests penetrate into every fibre of the social frame' (1999, 184), the scourge of industrial modernity threatened to transform even this 'fair world' into a 'scene of desolation' (170): it 'shuts out the pure light of heaven', he wrote, 'and defiles the very air [man] breathes' (170).

In the mind's eye of Wheeler's narrator, therefore, environmental (in)justice extends to the nonhuman, natural world. That world is, as Wheeler's narrative highlights, a source of value, and itself valuable; it is a unity that subsumes human doings. Tentative as this emphasis on the natural world's inherent value might be, it coincides with the more general drift of Chartist literature towards a discourse that challenged the contemporary narrative of environmental determinism, and its dualistic assumptions; as Chartist writers recognised, that narrative linked and legitimated the exploitation of the working classes and the nonhuman world by naturalising one and inferiorising the other. In the future that Wheeler and his fellow Chartists envisaged, a fair and just society would be one in which the constitutive links between people and place, human and nonhuman would be thought through in new and creative ways, beyond the material-discursive limits increasingly defined by industrial modernity.

*

As Chartist literature demonstrated through its very variety – and through its transgressive willingness to appropriate literary forms such as poetry and the realist novel – the working-class experience of environmental inequality resisted reductive generalisation. Even in extremis, these narratives suggested, working-class communities of care struggled against their own demoralisation, whilst simultaneously striving towards a reconfiguration of society that might challenge industrial modernity's triumphant march, and in so doing make room for an emergent, extensive form of justice. But what, then, was the fate of Chartism itself? What came of these complex attempts to re-write working-class identities, and in so doing, advance the cause of political participation? In 1848, the year of revolutions across Europe, Chartism petitioned government for the last time. As Wheeler glosses in *Sunshine and Shadow* (1999, 179–80), the Petition failed, in part because of the very problem that Chartist writers set out to address, with their countervailing narratives of a self-aware, self-educated, and law-abiding people: those in power denied the Petition's legitimacy, claiming that many of the signatures were false (Chase 2007, 312–3). The movement 'foundered not in blood, but in ridicule' (Halévy 1951 [1970], 240). Once broken, adds John Saville, Chartism 'was submerged in the national consciousness', lost 'beneath layers of false understanding and denigration' (1990, 202).

Yet less than two decades later, the government extended the vote to the majority of working men (1867). Arguably, two decades of reformist compromise across the social divide, assisted by years of relative prosperity and stability, were sufficient to persuade the bourgeoisie that, in the end, they had nothing to fear from working-class participation in the political system. But it might also be argued that these decades solidified a particular working-class identity which was itself immured to real or radical change. Chartism had failed, but union strength continued to grow. That strength did not translate into revolutionary fervour or acts of sedition. Marx and Engels had predicted revolution; instead, and as Max Weber would later argue, these manifestations of a divided society seemed to become a source of social stability (Thompson 2003, 93–4), as the unions used their power to negotiate better wages and better terms of employment with employers who were themselves anxious to avoid conflict (Lazonick 1994 [1999], 102). Power was shared, but in so doing, the aristocracy of labour itself became a conservative influence that forestalled 'the coming together of major class groups to challenge the status quo' (Thompson 2003, 93).

For many in the middle classes, the 1850s therefore marked the beginning of a new period of domestic peace and prosperity, symbolised by the success of the Great Exhibition in 1851. Not all were convinced, however, that Carlyle's 'Condition-of-England Question' had been addressed (Halévy 1951 [1970], 252–3). Indeed, Carlyle himself continued to exert a sometimes indefinable and idiosyncratic influence (Halévy 1951 [1970], 258), not least over writers such as Charles Dickens, whom Carlyle called 'an event world-wide; a *unique* of Talents' (italics in the original; Collins 1986 [1997], 203). Dickens was, as Chartists themselves acknowledged, 'difficult to ignore: not only was he a genius and spectacularly popular, but he also called attention to the same social issues that Chartists had raised' (Rose 2021, 36). It is, therefore, to Dickens' work that I turn in the next two chapters. As Robert Douglas-Fairhurst points out, 'disenchantment with political radicalism' may have become 'ever more ingrained in the public mood' (2021, 13) in the wake of the Chartist Petition's failure, but Dickens' own fiction was about to take a darker turn, reflecting 'his growing sense of a serious social mission, and his understanding of the kind of narrative that would be required to do it justice' (2021, 24). *Bleak House* (1852–3) is the 'first of these "dark" novels' (Douglas-Fairhurst 2021, 24). Here, Dickens offers his readers a sprawling vision of a chaotic modern society, and unpicks the connections that constitute it. In so doing, Dickens implicates his well-to-do readers in the existence of a slum he calls Tom-All-Alone's, representative product of a system of law that rewards itself rather than serving the cause of justice, but also of the corruption inherent in a self-interested society – the corollary of a political economy derived, as the historian G. M. Trevelyan memorably put it, from 'those parts of Adam Smith, [Thomas] Malthus and [David] Ricardo which suited the acquisition of wealth by the wealthy' (1937 [1969], 143). As I discuss, tracing the effects of that political economy is part of the novel's purpose, in an early expression of what would now be described as 'green criminology', or 'the exploration and examination of causes of and responses to "ecological", "environmental", or "green" crimes, harms and hazards' (Brisman and South 2020, 4–5) – a subfield within criminology which itself insists that the 'environment' is relevant to 'all legislation and [permeates] all facets of life' (4).

*

Notes

1 The point was well made by Ernest Jones, one of Chartism's later leaders, in a collection of tales entitled *Woman's Wrongs* (1855); here, one of the characters declare that '[i]t is folly to say "we can't help it", "we are creatures of circumstances" – "we are what society makes us". We *can* help it, we can *create circumstances*, we can *make society*' (Jones 2018, 125). Jones was referring to the question of women's political participation, but his message is entirely consistent with his work and views as a Chartist.

2 There have been determined efforts to rekindle interest in working-class and Chartist literature, pioneered by figures such as Martha Vicinus (*The Industrial Muse: A Study of Nineteenth-Century British Working-Class Literature*, 1974), H. Gustav Klaus (who edited *The Socialist Novel: Towards the Recovery of a Tradition*, 1982, and wrote *The Literature of Labour*, 1985), Peter Schekner (*An Anthology of Chartist Poetry: Poetry of the British Working Class, 1830s–1850s*, 1989), and, more recently, Ian Haywood, who has written about Chartist fiction (1997) and edited three volumes of Chartist fiction (1995, 1999, 2001; the last of these used in its reprinted, 2018 form); Haywood has also surveyed Chartist literature and its subsequent reception in 'The Literature of Chartism' (2016). For a more detailed discussion of the literary history of Chartist poetry, see Mike Sanders (*The Poetry of Chartism*, 2009, 38–68), who also touches on the more general emergence of critical interest in Chartist and working-class literature (42–4). As a result of these literary efforts to recover a Chartist literary tradition, at least some of that literature is now available to the general reader, whether in print anthologies or in the open-access digital domain; there is, furthermore, a website source, Chartist Fiction Online (https://www.chartistfiction.hosting.nyu.edu/), hosting bibliographic records of various literary works and reviews published in Chartist periodicals.

3 Reports on the condition of the Leicester stockingers sometimes convey just this sense of powerless at language's limits: R. A. McKinley quotes from one investigation into contemporary conditions, which noted that 'work was scarce and wages were so low that it hardly paid to be at work at all … Misery and want were stamped on all their ["the stockingers"] careworn and anxious features, and the wretchedness was too severe *to be portrayed*' (italics added; 1958, n.p.).

4 Drawing on the biophilia hypothesis posited by Edward O. Wilson (1984), in which Wilson refers to an 'innate tendency to focus on life and lifelike processes' (1984 [2003], 1), Ellen Miles argues that this 'need for nature' is 'genetically encoded' (Miles 2022, 17); by extension, contact with the natural world is 'not a luxury or a frivolity – nature is a human right' (18). To the contrary, Simon Estok argues, controversially, humankind's 'natural' condition is ecophobic; we are, in fact, encoded with an 'antipathy towards nature' that might once have helped 'preserve our species' but now fosters the kind of 'growth economics and ideological interests' inimical to sustainability (Estok 2018, 1). New materialism offers no answers to this polarised debate, but its emphasis on material-discursive formation does suggest that regardless of the nature of human responsiveness to the nonhuman and natural, it is always and invariably mediated by discursive constructions, amongst them the kind of Romanticised, beneficent 'Nature' that Jones invokes in his poem.

Works cited

Attridge, Derek. 2004. *The Singularity of Literature*. London: Routledge.

Beck, Ulrich. 1992 [2012]. *Risk Society: Towards a New Modernity*, translated by Mark Ritter. London: Sage.

Beck, Ulrich. 1995 [2002]. *Ecological Politics in an Age of Risk*, translated by Amos Weisz. Cambridge: Polity Press.

Breton, Rob. 2013. 'Diverting the Drunkard's Path: Chartist Temperance Narratives'. *Victorian Literature and Culture* 41 (1): 139–52. Available at jstor.org, www.jstor.org/stable/24575676. Accessed 29 March 2022.

Brisman, Avi, and Nigel South 2020. 'Introduction: New horizons, ongoing and emerging issues and relationships in green criminology'. *Routledge International*

Handbook of Green Criminology, 2nd edn. Edited by Avi Brisman and Nigel South, 1–36. London: Routledge.

Buell, Lawrence. 2001. *Writing for an Endangered World: Literature, Culture, and Environment in the U.S. and Beyond*. London: Belknap Press.

Carlyle, Thomas. 1839 [1897]. *Chartism. The Works of Thomas Carlyle, Volume Sixteen*, 204–85. New York, NY: Peter Fenelon Collier. Available at archive.org, https://archive.org/details/worksofthomascar0016carl/page/n7/mode/2up?view=theater. Accessed 30 November 2022.

Chase, Malcolm. 2007. *Chartism: A New History*. Manchester: Manchester University Press.

Chinn, Carl. 1995 [2006]. *Poverty amidst Prosperity: The Urban Poor in England, 1834–1914*. Lancaster: Carnegie.

Clapp, B. W. 1994. *An Environmental History of Britain since the Industrial Revolution*. London: Longman.

Clark, Anna. 1992. 'The Rhetoric of Chartist Domesticity: Gender, Language, and Class in the 1830s and 1840s'. *Journal of British Studies* 31 (1): 62–88. Available at jstor.org, www.jstor.org/stable/175877. Accessed 29 March 2022.

Clapp, B. W. 1994. *An Environmental History of Britain since the Industrial Revolution*. London: Longman.

Collins, Philip, ed. 1986 [1997]. *Dickens: The Critical Heritage*. Abingdon: Routledge.

Cook , Sir Edward. 1913. *The Life of* Florence *Nightingale*, *Volume 1 (1820–1861)*. London: Macmillan. Available at archive.org, https://archive.org/details/lifeflorencenigh01cook/page/n7/mode/2up. Accessed 14 March 2022.

Coolsaet, Brendan, and Pierre-Yves Néron. 2021. 'Recognition and Environmental Justice'. *Environmental Justice: Key Issues*, edited by Brendan Coolsaet, 52–63. London: Routledge.

Cooper, Thomas. 1845. *Wise Saws and Modern Instances*, *Volumes* I *and* II. London: Spottiswoode for Jeremiah How. Available at archive.org, https://archive.org/details/wisesawsmodernin01coop/page/n281/mode/2up. Accessed 18 March 2022.

Daunton, Martin. 2000. 'Society and Economic Life'. *The Nineteenth Century*, edited by Colin Matthew, 41–82. Oxford: Oxford University Press.

Davenport, Romola J. 2020. 'Urbanisation and Mortality in Britain c.1800–1850', *Economic History Review* 73 (2): 445–85. Doi: 10.1111/ehr.12964

Douglas-Fairhurst, Robert. 2021. *The Turning Point: A Year That Changed Dickens and the World*. London: Jonathan Cape.

Eagleton, Mary, and David Pierce. 1979. *Attitudes to Class in the English Novel*. London: Thames and Hudson.

Estok, Simon C. 2018. *The Ecophobia Hypothesis*. London: Routledge.

Engels, Friedrich. 1993 [2009]. *The Condition of the Working Class in England*, edited and introduced by David McLellan. Oxford: Oxford University Press.

Foot, Paul. 2005 [2006]. *The Vote: How It was Won and How It was Undermined*. London: Penguin.

Gallagher, Catherine. 1985. *The Industrial Reformation of English Fiction: Social Discourse and Narrative Form, 1832–1867*. Chicago, IL: University of Chicago Press.

Garrard, Greg. 2012. *Ecocriticism*. 2nd edn. Abingdon: Routledge.

Gaskell, Elizabeth. 1848 [2003]. *Mary Barton: A Tale of Manchester Life*, edited by Macdonald Daly. London: Penguin.

Gaskell, Elizabeth. 1854–5 [2000]. *North and South*, edited by Patricia Ingham. London: Penguin.
Gilbert, Pamela K. 2009. 'History and Its Ends in Chartist Epic'. *Victorian Literature and Culture* 37 (1): 27–42. Available at jstor.org, www.jstor.org/stable/40347212. Accessed 30 March 2022.
Gould, Peter C. 1988. *Early Green Politics: Back to Nature, Back to the Land, and Socialism in Britain, 1880–1900*. Sussex: Harvester Press.
Halévy, Élie. 1951 [1970]. *A History of the English People in the Nineteenth Century, Volume IV: Victorian Years, 1841–1895*. London: Ernest Benn.
Hardy, Dennis. 1979. *Alternative Communities in Nineteenth Century England*. London: Longman.
Haywood, Ian. 1995a. 'Introduction'. *The Literature of Struggle: An Anthology of Chartist Fiction*, edited by Ian Haywood, 1–25. Aldershot: Scolar Press.
Haywood, Ian, ed. 1995b. *The Literature of Struggle: An Anthology of Chartist Fiction*. Aldershot: Scolar Press.
Haywood, Ian. 1997. *Working-Class Fiction from Chartism to Trainspotting*. Plymouth: Northcote House.
Haywood, Ian. 1999. 'General Introduction'. *Chartist Fiction: Volume One*, edited by Ian Haywood, ix–xv. Aldershot: Ashgate.
Haywood, Ian. 2000. ''Graphic Narratives and Discoveries of Horror': The Feminization of Labour in Nineteenth-century Radical Fiction'. *British Industrial Fictions*, edited by H. Gustav Klaus and Stephen Knight, 5–23. Cardiff: University of Wales Press.
Haywood, Ian. 2016. 'The Literature of Chartism'. *The Oxford Handbook of Victorian Literary Culture*, edited by Juliet John, 83–102. Oxford: Oxford University Press.
Hilton, Boyd. 2006. *A Mad, Bad, and Dangerous People? England 1783–1846*. Oxford: Clarendon Press.
Jones, Ernest. 1855. *The Battle-Day, and other poems*. London: Routledge. Available at archive.org, https://archive.org/details/battledayandoth00jonegoog/page/n4/mode/2up?view=theater. Accessed 24 March 2022.
Jones, Ernest. 2018. *Woman's Wrongs*. Reprinted in *Chartist Fiction: Volume Two*, edited by Ian Haywood. London: Routledge.
Langland, Elizabeth. 2001. 'Women's Writing and the Domestic Sphere'. *Women and Literature in Britain, 1800–1900*, edited by Joanne Shattock, 119–41. Cambridge: Cambridge University Press.
Lazonick, William. 1994 [1999]. 'Employment Relations in Manufacturing and International Competition'. *The Economic History of Britain Since 1700, Volume 2: 1860–1939*, edited by Roderick Floud and Deirdre McCloskey, 90–116. 2nd edn. Cambridge: Cambridge University Press.
Ledger, Sally. 2002. 'Chartist Aesthetics in the Mid Nineteenth Century: Ernest Jones, a Novelist of the People'. *Nineteenth-Century Literature* 57 (1): 31–63. doi: 10.1525/ncl.2002.57.1.31
Lewontin, Richard, and Richard Levins. 2007. *Biology Under the Influence: Dialectical Essays on Ecology, Agriculture, and Health*. New York, NY: Monthly Review Press.
Luckin, Bill. 2015. *Death and Survival in Urban Britain: Disease, Pollution and Environment 1800–1950*. London: Bloomsbury Academic.
McDonagh, Josephine, 2021. *Literature in a Time of Migration: British Fiction and the Movement of People, 1815–1876*. Oxford: Oxford University Press.

McKinley, R. A., ed. 1958. *A History of the County of Leicester*: *Volume 4*, the City of Leicester, pp. 303–14. London: Victoria County History. Available at British History Online, www.british-history.ac.uk/vch/leics/vol4/pp303-314. Accessed 18 March 2022.

Messner, Andrew. 1999. 'Land, Leadership, Culture, and Emigration: Some Problems in Chartist Historiography'. *The Historical Journal* 42 (4): 1093–1109. Available at jstor.org, www.jstor.org/stable/3020938. Accessed 29 March 2022.

Miles, Ellen. 2022. 'Introduction'. *Nature is a Human Right*, edited by Ellen Miles, 14–24. London: Dorling Kindersley.

Morris, Pam. 2003 [2011]. *Realism*. London: Routledge.

Nasmyth, James. 1883. *James Nasmyth, Engineer: An Autobiography*, edited by Samuel Smiles. New York, NY: Harper & Brothers. Available at archive.org, https://archive.org/details/jamesnasmytheng00nasmgoog/page/n9/mode/2up. Accessed 20 February 2021.

Nixon, Rob. 2011. *Slow Violence and the Environmentalism of the Poor*. Cambridge, MA: Harvard University Press.

Nussbaum, Martha C. 2006. *Frontiers of Justice: Disability, Nationalities, Species Membership*. Cambridge, MA: Belknap Press.

Plumwood, Val. 1993 [1997]. *Feminism and the Mastery of Nature*. London: Routledge.

Rose, Jonathan. 2021. *The Intellectual Life of the British Working Classes*. 3rd edn. London: Yale University Press.

Sanders, Mike. 2000. 'Accidents of Production: Industrialism and the Worker's Body in Early Victorian Fiction'. *British Industrial Fictions*, edited by H. Gustav Klaus and Stephen Knight, 24–35. Cardiff: University of Wales Press.

Sanders, Mike. 2009. *The Poetry of Chartism: Aesthetics, Politics, History*. Cambridge: Cambridge University Press.

Saville John, . 1990. *1848: The British State and the Chartist Movement*. Cambridge: Cambridge University Press.

Taylor, Jesse Oak. 2016. *The Sky of Our Manufacture: The London Fog in British Fiction from Dickens to Woolf*. London: University of Virginia Press.

Thompson, Dorothy. 2015. *The Dignity of Chartism*, edited by Stephen Roberts. London: Verso.

Thompson, Neil. 2003. *Promoting Equality: Challenging Discrimination and Oppression*. 2nd edn. Houndmills: Palgrave Macmillan.

Trevelyan, George Macaulay. 1937 [1969]. *British History in the 19th Century and After: 1782–1919*. 2nd edn. London: Longmans, Green & Co.

Vanden Bossche, Chris R. 2014. *Reform Acts: Chartism, Social Agency, and the Victorian Novel, 1832–1867*. Baltimore, MD: John Hopkins University Press.

Vargo, Gregory. 2011. 'A Life in Fragments: Thomas Cooper's Chartist 'Bildungsroman''. *Victorian Literature and Culture* 39 (1): 167–81. Available at jstor.org, www.jstor.org/stable/41307856. Accessed 18 March 2022.

Voth, Hans-Joachim. 2004. 'Living Standards and the Urban Environment'. *The Cambridge Economic History of Modern Britain, Volume 1: Industrialisation*, 1700–1860, edited by Roderick Floud and Paul Johnson, 268–94. Cambridge: Cambridge University Press.

Walton, Samantha. 2021. *Everybody Needs Beauty*: *In Search of the Nature Cure*. London: Bloomsbury.

Ward, J. T. 1970. *The Factory System*, *Volume II*. Newton Abbot: David & Charles.

Watt, Ian. 1957 [2000]. *The Rise of the Novel*. London: Pimlico.

Wheeler, Thomas Martin. 1999. *Sunshine and Shadow: A Tale of the Nineteenth Century*. Reprinted in *Chartist Fiction: Volume One*, edited by Ian Haywood, 72–193. Aldershot: Ashgate.

Williams, Raymond. 1958 [1959]. *Culture and Society, 1780–1950*. London: Chatto & Windus.

Wilson, Edward O. 1984 [2003]. *Biophilia*. Cambridge, MA: Harvard University Press.

4 Seeking justice in Charles Dickens's *Bleak House*

First published in 1852–3, *Bleak House* is perhaps the greatest of Dickens's commentaries on the Carlylean 'Condition-of-England Question' (discussed in Chapter 1), and on a growing social, economic, and environmental divide that impacted on the poor but implicated the rich. 'In writing *Bleak House*', notes J. Hillis Miller, Dickens therefore 'constructed a model in little of English society in his time' (1991, 179), 'a miniature version of the interconnectedness of people in all levels of society' (180). In turn, the novel offers one of the period's most powerful portrayals of environmental injustice, in the form of a slum district called Tom-all-Alone's.[1]

Like the real-life 'rookeries' on which it is modelled, Tom-all-Alone's is the last refuge of what one contemporary described as 'the pariahs, so to speak, of the body social' (Beames 1852, 4). Its relationship to the forces that were then transforming the nation is, however, a complex one. Tom-all-Alone's is not, like the working-class districts of Coketown in *Hard Times* (discussed in Chapter 5) or the wastelands of Belle-Isle (discussed in the Introduction), an obvious product of industrial modernity; it is not tainted by 'the incense of some foul factory [or] gas-works' (Dyos 1967, 25). Embedded within the decaying fabric of one of London's oldest districts, it is the product of an unreformed legal system, whose self-serving obfuscations ensure that contested estates are discussed and debated for years and even decades, whilst any property caught up in them slowly falls to rack and ruin. Yet there is a link between the environmental injustice that Dickens depicts, and society's wider transformation. By the middle of the nineteenth century, industrialisation and urbanisation had become interlocked features of industrial modernity, the one requiring the other, and this was no less true in London, which was an important industrial centre in its own right. As the city's population grew, the working classes sought out whatever accommodation they could find; the most desperate, like the brick-makers whom Dickens depicts at various stages throughout the novel, settled in London's rookeries. In turn, their concentration in these squalid corners of the city made them more vulnerable to illness and disease. It was not industrial toxins that destroyed their lives – although the city's smoky skies played their part – but epidemics generated by a basic lack of sanitation. Society, meanwhile, carried on obliviously, in a powerful demonstration of the psycho-geographical separation of rich and

DOI: 10.4324/9780367821609-5

poor that made environmental injustice possible amidst plenty. This was not just a failure to recognise the plight of the poor, but a failure to acknowledge society's involvement in and responsibility for their plight.

This is Dickens's subject, in a novel whose very comprehensiveness enables him to trace the complex interactions that connect the unlikeliest figures. In so doing, Dickens has something profound to say about the nature and importance of relationships in mass, modern societies. The complexities and confusions of new ways of living require new ways of understanding, which in turn require new forms of ethical consideration. As the agentiality of reality becomes apparent, emerging risks highlight the co-constitutive nature of human/ nonhuman entanglements, on which all forms of flourishing depend. As Juliet John insists, Dickens always 'views things' in a 'relational way ... never existing in isolation' (John 2012, 117; see also Lesiuk 2018, 121). What matters amidst relationality, Dickens's novel suggests, is the quality of those relationships, relationships that transcend the myriad boundaries and dualistic divides that society's deeply entrenched 'logic of colonisation' (Plumwood 1993 [1997], 43) has established. It is in these terms, as I discuss in more detail, below, that we should read Dickens's memorable portrayal of Tom-all-Alone's. As his eloquent and inventive prose underlines, it is 'a ruinous place' (Dickens 2003, Ch. 16, 256)[2] in which poverty intersects with squalor, and the poor are everywhere exposed to environmental harms and denied its goods (fresh air, clean water, even a glimpse of blue skies). But as Dickens insists, Tom-all-Alone's problems are intimately bound up with the workings of society itself, and isolated as it may seem from the everyday world, it still has the power to reach out and devastate the lives of the more fortunate.

*

In Chapter 22 of *Bleak House*, a detective named Bucket sets off in search of a street urchin known only as Jo. To assist him in the search, he engages the assistance of Snagsby, a mild-mannered, respectable law-stationer. Snagsby knows poor Jo, and is never slow to slip him a charitable half-crown. Together, the two men set off to visit the London slum where Jo is believed to live. This is Tom-all-Alone's, and Snagsby is appalled by what he finds there.

> Mr. Snagsby passes along the middle of a villainous street, undrained, unventilated, deep in black mud and corrupt water – though the roads are dry elsewhere – and reeking with such smells and sights that he, who has lived in London all his life, can scarce believe his senses. Branching from this street and its heaps of ruins are other streets and courts so infamous that Mr. Snagsby sickens in body and mind and feels as if he were going every moment deeper down into the infernal gulf.
>
> (Ch. 22, 358)

And so Snagsby stumbles into the abyss that poor Jo calls home.

As Lawrence Buell explains, the image of a descent into an underworld is a staple of nineteenth-century exposés of slum conditions, and it was calculated to 'instil shock and compassion in uninitiated readers' (2001, 43). It functions in two ways, by allegorising the slum 'in classico-biblical terms', as hellish underworld, and by situating it within a Gothicised depiction of the modern city as a terrifying labyrinth, everywhere concealing unseen horrors (Buell 2001, 43). As poor Snagsby discovers, 'the real locale of Gothic horror' (Pritchard 1991, 437) is there in the very heart of the metropolis, just yards from his own place of work, in streets where 'mud, ooze, and excrement become blurred' (Lougy 2002, 478).[3]

Snagsby is appalled, no doubt because he believes that (as Inspector Bucket observes 'in a comfortable and soothing manner') 'every person should have their rights according to justice' (Ch. 22, 356). Whether or not this is what Bucket wants is another matter, since he 'is allowed to hound [Jo] with impunity' (Allan 2003, 17) – indeed, 'Bucket's insistence that Jo simply "move on" from place to place illustrates a larger political failure to develop reform strategies for the urban poor' (Joyce 2002, 120) – but Snagsby certainly believes in doing the right thing, and endeavours to do it. The sights and sounds that he encounters in Tom-all-Alone's are, however, of a scale and a nature that leave him dumbfounded and disbelieving, and Snagsby

> goes homeward so confused by the events of the evening that he is doubtful of his being awake and out – doubtful of the reality of the streets through which he goes – doubtful of the reality of the moon that shines above him.
>
> (Ch. 22, 365)

Yet slums like Tom-all-Alone's were real enough, and Dickens's intention was to impress as much on his readers. Still, there is a deeper issue at stake in *Bleak House*, encapsulated by the fog with which the novel opens so memorably. 'Fog everywhere', observes the omniscient narrator, '[f]og up the river, where it flows among green aits and meadows; fog down the river, where it rolls defiled among the tiers of shipping and the waterside pollutions of a great (and dirty) city' (Ch. 1, 13, 14). Fog everywhere – but much of it is smoke. 'Smoke lowering down from chimney-pots, making a soft black drizzle, with flakes of soot in it as big as full-grown snowflakes – gone into mourning, one might imagine, for the death of the sun' (Ch. 1, 13). This is no longer fog, to which London was in any case susceptible, but an instance of what Jesse Oak Taylor has described as the abnatural, a term that marks both the withdrawal of the natural, and its persistence in modified new forms (2016, 5). The 'London Particular' (Ch. 3, 42) that the narrator describes is what would later be called 'smog', an early twentieth-century conflation of 'smoke' and 'fog', but already, as here, a nineteenth-century reality.[4] London's smog has its own long history, as Peter Brimblecombe has discussed (1987, 63–89), but the problem of it increased markedly as the city industrialised, and factories added their own plumes of smoke to those created by domestic coal consumption. '[S]moke-producing monopolists' (as Lord Palmerston contemptuously described the industrialists of the age) could and did claim that the smoke nuisance was as much a domestic as it was an industrial problem (*Hansard* HC Deb., 8 August 1853), but the two were interlinked: as cities industrialised, so they drew in labour, adding to population growth and hence domestic coal use. This was true not just of the 'shock' cities of the industrial heartlands (like the Manchester that Engels described so vividly) (Briggs 1963 [1987], 56), but of long-established centres like London itself. The smog that fills the city's streets is, therefore, the very real, very toxic product of a world warmed but also powered by fossil-fuel, and transformed by industrial modernity.

In the world of the novel, however, that smog also stands for a complex social disorder whose overlooked but iniquitous implications Dickens sets out to lay bare, and it relates to the astonishing scale and rate of urban growth. This was most obvious in prominent industrial centres such as Manchester, which quadrupled in size between 1801 and 1851, but London's population itself doubled in the same short period (Hilton 2006, 6). The growth in population was matched by a growth in spatial extent, as cities sprawled with a chaotic speed that defied comprehension, adding to the impression that these places were at best mysterious, and at worst dangerous and threatening (Hoppen 1998, 52). For their citizens, many first-generation arrivals, the result was a daunting sense that cities were in some profound sense unknowable, and relations within them inconceivable; that, whatever the real and essential connections that continued to define human society, they were 'obscured, complicated, mystified' (Williams 1973 [1975], 155) by a smoke-wreathed, fog-bound new reality.

In spite of their continuing and inevitable interdependence, the characters in Dickens's fiction therefore see themselves as isolated individuals, forced to make their own way, without the benefit of company or community; to suffer, and sometimes go mad in isolation. It takes an Esther Summerson to bridge this gulf – and as the novel's principal character, her first-person narrative takes up a substantial proportion of it. Esther is not, however, the only agency who embodies and asserts this interdependency, even if she may well be the only positive instance of it in the novel. Other forces and flows are at work, and they are negative ones, life-threatening and corrosive. Esther will fall victim to one of them even as she tries to counter this growing trend towards social isolation and mutual impoverishment. 'Viruses are radical democrats', notes Eagleton, 'disdainful of social distinction' (2005, 161); and in *Bleak House*, the focal point of the city's infection is, inevitably, Tom-all-Alone's, whose name itself suggests urban alienation (Eagleton 2005, 160).

Like Jacob's Island in Dickens's earlier novel, *Oliver Twist* (1838), Tom-all-Alone's is a characteristic example of what was then known as a rookery, whose depiction would (Dickens hoped) help to open the eyes of those who were 'purposely blind' to the realities of working-class life (Dickens 1854 [1998], 226). 'There is much in a name', wrote the Reverend Thomas Beames in *The Rookeries of London*, 'one significant phrase which spares circumlocution' (1852, 1): these slums, it was thought, resembled noisy, quarrelsome rook colonies, 'the houses for the most part high and narrow, the largest possible number crowded together in a given space ... all descended to the lowest scale which is compatible with human life' (2). The rook was once regarded as an agricultural pest (Cocker 2007 [2016], 130), but it quickly took to scavenging in cities; appropriately dressed in black, the rook was also, moreover, a bird of ill-omen (Cocker 2007 [2016], 133), and these (largely negative) associations perhaps explain why the term rookery became 'the current money of the realm of thought' (Beames 1852, 1).

There was, however, a still more negative association at work in this process of naming. The work 'rook' carries a secondary meaning, to steal or cheat, which overlapped with the prevalent view that slums such as these were 'rendezvous of vice, whose effects we feel in street robberies and deeds of crime ... *nurseries* of felons' (Beames 1852, 149). They are '[h]uge chimeras of crime', wrote Beames, 'embracing in their outstretched arms districts not yet polluted' (1852, 150). Vice was not the only anxiety which the rookeries provoked in middle-class readers, and on which reformers like Beames played. Rookeries were, as Beames reminded his readers, 'beds of pestilence, where the fever is generated which shall be propagated to distant parts of the town' (1852, 149). These concerns were often genuine. As Romola J. Davenport notes, '[h]igh population densities favoured [disease] transmission'; moreover, 'most cities provided inadequate facilities for the disposal of the volumes of wastes generated by such densities and numbers of humans and animals, and for the prevention and treatment of gastrointestinal diseases associated with these living conditions' (2020, 455).

From our own, late-modern perspective, the term 'rookery' also highlights its abnatural identity (Taylor 2016, 5), as a place from which nature is in fact excluded, or present only in lethal forms (such as disease), or as toxic hybrids such as the London particular, which helped obscure its very existence. As we would also now recognise, the rookery is a defining example of environmental injustice, as a site where 'human suffering and social inequality' converge (Pellow 2018, 2), and one 'particular social group is disproportionately affected by environmental hazards' (5). But behind that injustice lies another, related process. Justice requires awareness not only of hazards and harms, but recognition of those affected by them; as Brendan Coolsaet and Pierre-Yves Néron point out, recognition is an essential prerequisite for political participation, without which the unfair distribution of harms and goods cannot be addressed (2021, 52–3). In Victorian Britain, however, the problem was not (or not simply) that society rendered the slum dwellers invisible. As we have already seen, the problem was one of 'mis- or malrecognition' as itself 'a cultural and institutional form of injustice' (Schlosberg 2007, 16), whereby society actively constructs a distorted image of those it disadvantages, and so excuses its own failures in relation to them. What is noticeable about Victorian society's responses to slums such as the rookeries, notes Gertrude Himmelfarb, is that 'the same words – "residuum", "refuse", "offal" – were used to denote the sanitary problem and the human waste that constituted the human problem' (1984, 358). Increasingly, those at the lowest end of the social scale were themselves categorised as society's irremediable residue. As Carl Chinn explains, contemporaries distinguished this urban 'residuum' from the labouring classes, and labelled them as 'the ragged', a degenerate and dangerous underclass who could work, but refused to do so (1995 [2006], 94–5): thus, '[o]bservers robbed the so-called "ragged" of their membership of the working class as they stole from them their humanity' (1995 [2006], 95). Karl Marx himself described them as 'social scum', a 'passively rotting mass' (qtd in Himmelfarb, 1984, 378), and concluded, adds Chinn, that they were 'potentially counter-revolutionary' (1995 [2006], 95). There was, furthermore, a racial dimension to this demonisation of the 'ragged', linked to Irish immigrants, by then a substantial presence in London: 'wherever in London has expressively been called a "*Rookery*" exists', wrote the Reverend John Garwood in 1853, 'we may be sure that it is inhabited by Irish. Our English poor will not live in that manner' (italics in the original; 1853, 314).

If these were 'rookeries', therefore, it was because their occupants were themselves less than human, and as such, undeserving. Conveniently, this logic converged with the *laissez-faire* refusal to engage the state in solving the problem that rookeries represented, and the free market had engendered, as places

> where, by the unwritten laws of competition, rents rise and room is economised in proportion; where, because there is no restraint to check the progress of avarice, no statute to make men do their duty, they turn to profit the necessities of their fellow-creatures, and riot on the unhallowed gains which injustice has amassed at the expense of the poor.
>
> (Beames 1852, 2)

For those like Beames, the continued existence of rookeries amidst so much prosperity was a source of shame, and an iniquity that must be challenged; but how? The discursive construction of the rookery's inhabitants as inconsequential 'human waste' – and as members of a refractory and intransigent underclass – provided society with the excuse to ignore their predicament.

What discourse had engendered, however, it might yet undo. To the well-intentioned, like the hopeful Beames, works such as his own had a role to play in altering these views, and in so doing, materially improving living conditions for the poor. 'Mr. Dickens' had himself set the example in *Oliver Twist*, wrote Beames, by dragging the rookeries 'into light' (1852, 52). The difficulty lay in finding a way in which to command middle-class interest: Beames relied quite as much on stoking self-interested middle-class fears of violence, vice, and disease as he did on appealing to its better instincts, or by returning to the 'residuum' its humanity, and enabling this stereotyped mass to be seen as a succession of individual stories and subjectivities, united only by 'common necessity' (Beames 1852, 2). Nor was Dickens averse to these tactics. In *Dombey and Son* (1846–8), a novel published several years before *Bleak House*, he appealed for

> a good spirit who would take the housetops off … and show a Christian people what dark shapes issue from amidst their homes, to swell the retinue of the Destroying Angel as he moves forth among them! For only one night's view of the pale phantoms rising from the scenes of our too-long neglect; and from the thick and sullen air where Vice and Fever propagate together, raining the tremendous social retributions which are ever pouring down, and ever coming thicker!
>
> (Dickens 2002, 702)

But as Dickens continues his narrative, it becomes clear that he is not simply making an appeal to middle-class self-interest, by dangling before them the threat of disease or crime. He is laying down a challenge to the 'contracted sympathies and estimates' of his contemporaries, and seeking their understanding and their sympathy; he is reminding them that they are a part of a larger lifeworld in which they are all intimately involved and for which they too bear responsibility, 'owing one duty to the Father of one family, and tending to one common end, to make the world a better place!' (Dickens 2002, 702). The key, suggests the narrator of *Dombey and Son*, lies in rousing this 'Christian people' to look 'out upon the world of human life around them, to a knowledge of their own relation to it' (Dickens 2002, 702).

'This creation of consciousness – of recognitions and relationships – can then be seen as the purpose of Dickens's developed vision', argued Raymond Williams (1973 [1975], 155), a vision that itself coincides with the environmental justice movement's emphasis on recognition as the basis for participation. As Caroline Levine adds, this vision achieves its fullest working out in *Bleak House*, a novel whose form enacts this 'creation of consciousness' by setting out to represent nothing less than 'the social life of the nation' (2012 [2013], 91). *Bleak House* embodies the problem that Dickens set out to address: at once a history and a herstory – there are two narrators, one implicitly male, the other explicitly female – the novel offers a vast, sprawling narrative spread over 67 chapters, with competing and seemingly unconnected plots lines, numerous settings both urban and rural, and 'more than seventy idiosyncratic characters' (Levine 2012 [2013], 91) of such decided peculiarity and from such widely differing parts of society that it seems nothing can possibly connect them (92). What relationships could there be, the novel asks as its focus alights on Tom-all-Alone's and the 'swarm of misery' that inhabits it (Ch. 16, 256), between these various people and places? 'What connexion', the narrator replies, 'can there have been between many people in the innumerable histories of this world, who, from opposite sides of great gulfs, have nevertheless, been very curiously brought together!' (Ch. 16, 256).

The challenge that the novel embodies in its very form is, therefore, the challenge of Victorian society as a whole, grown so fast and become so fluid that it now seems unknowable and threatening – a truly Gothicised environment, marked out by 'secrecy, alienation and monstrosity' (Willis 2012 [442014], 18). It is also the challenge posed by the environmental iniquities embodied in slums such as Tom-all-Alone's, for which nothing and no one seems responsible: all these links and connections have been obscured, lost in the enveloping fog. In the atomised and fragmented society that is the result, the isolated individual stands and falls alone, like the 'tumbling tenements' of Dickens's rookery (Ch. 16. 256). 'The reality is that we are members of one another', argues Terry Eagleton; 'the problem is that we are bound up with each other in a way which conceals the fact from our consciousness' (2005, 161). To return to the question begged by the very existence of Tom-all-Alone's, how can we discover who or what is responsible for the iniquities that distort society? Dickens's answer, as Williams suggested, is to insist on the reality of relationality; to confront his readers with the truth of lived experience in the slums, and force them to recognise their own links to it; to insist that society must take responsibility for its own. We are all entangled with the lives of others, a fact central to the converging narrative arcs that *Bleak House* describes, and central to the nature of the sometimes threatening, Gothicised world that Dickens constructs in the novel.

It is a world, as we might now recognise, that closely resembles the new materialist reconceptualisation of it. Simply, new materialism emphasises humankind's constitutive entanglement with an agential reality: as Karen Barad insists, we are not in any final sense separable from this 'dynamic and shifting entanglement of relations' (Barad 2007, 35), and we cannot therefore draw a sharp dividing line between ourselves and the world.[5] This is Dickens's own starting point: he begins by rejecting the separation of subject and object, and with it, the process of radical exclusion (Plumwood 1993 [1997], 49) that distinguishes and then prioritises one over the other. As Val Plumwood has argued, this act of separation is the basis of a 'logic of colonisation' (1993 [1997], 41) that naturalises the construction of 'others' as a homogenous and inferior group (52–3). In a further extension of this logic of domination, the 'inferiorised group' (Plumwood 1993 [1997], 53) is instrumentalised and objectified; defined in terms of the role they perform or the job they do, and stripped of intrinsic value, they are simply regarded as a resource (53). Like the residuum that inhabited the wastelands of London's rookeries, people come to seem less than human; they come to resemble things or objects. But in Dickens's world, nothing ever is *merely* a thing or *simply* an object. Indeed, matter is itself alive and lively, in unexpected, unpredictable, and sometimes unwelcome ways.

As John Parham points out, Dickens possessed a 'proto-ecological … attentiveness to matter' (2017 [2019], 119), and this carries over into the kind of realist mode Dickens adopts in *Bleak House*: when (in the novel's Preface) he promised to dwell 'upon the romantic side of familiar things' ('Preface', 7), he was in effect promising to bring to bear a visionary imagination that would offer, instead of a mechanistic literalism, a 'fantastic fidelity' (qtd in Forster n.d., 562) – a heightened appreciation – of a dynamic and lively world. The result, notes Juliet John, 'is a cliché of Dickens criticism': that his writing is 'thus a form of thing art' (2012, 130). As the historian Asa Briggs remarked, in a landmark account of *Victorian Things*, Dickens 'could even write a story in *Household Words* about a talking hat-stand, "my mahogany friend"' (2003, 20).

In Dickens's fictional world, so-called 'objects' are not just important for what they might signify or symbolise. They are meaningful in themselves, and in themselves generative; they possess what Jane Bennett describes as 'thing power, the material agency of natural bodies and technological artefacts' (2010, xiii). They act, even if they act only through deconstruction, as the ruinous houses that make up Tom-all-Alone's do with frightening regularity. It follows that things are no longer merely things, passive and pliable. They demand our attention and our consideration. At every scale, whether unimaginably tiny or vast, and over timescales that may be inconceivably small or large, objects have effects – and effects on us. It also follows that things have their own identities, their own characteristics. Some entities appear friendly, like the eponymous Bleak House, a place which gives Esther her first glimpse of what a safe, settled life might be like; to Esther, the various objects with which it is filled speak of continuity and contentment, and help make it a place of 'light, and warmth, and comfort' (Ch. 6, 87; see Tait 2020, 124–5). Other things are, however, inimical or indifferent. Some seem frightening, threatening; they may even constitute the risk in a Victorian risk society.

It is here that Dickens once again draws on the Gothic, in support of a 'complex epistemology that figures the real as the multiple convergences of natural and unnatural, ordinary and extraordinary' (Willis 2012 [2014], 18). As Ulrich Beck has argued, modernisation exceeds and therefore breaks '*science's monopoly on rationality*' (italics in the original; Beck 1992 [2012], 29): it defies expectations, and in turn Dickens's use of Gothic marks the limits of what society can comprehend, or cram into its own definition of what constitutes (a passive, malleable) reality. To the contrary, reality is active: as Bennett adds, the term 'thing power' marks 'the strange ability of ordinary … items to exceed their status as objects and to manifest traces of independence or aliveness, constituting the outside of our own experience' (2010, xvi). Thus, Tom-all-Alone's embodies the 'common tropes of entrapment and decay that typify Gothic architecture' (Willis 2012 [2014], 19) – and the term 'rookery' is perfectly suggestive of a whirl of corvids, haunting this 'already Gothicised space' (19) – but the way in which it transforms itself through decay is itself memorable, and monstrous:

> Twice lately there has been a crash and a cloud of dust, like the springing of a mine, in Tom-all-Alone's; and each time a house has fallen. These accidents have made a paragraph in the newspapers and have filled a bed or two in the nearest hospital. The gaps remain, and there are not unpopular lodgings among the rubbish. As several more houses are nearly ready to go, the next crash in Tom-all-Alone's may be expected to be a good one.
>
> (Ch. 16, 257)

Dickens's tone is ironic: when poverty is so absolute, even collapsed buildings provide 'not unpopular' lodgings; but what he describes is 'an Abbau, or unbuilding', in the words of Lewis Mumford, a kind of evolution through entropy (1961 [1973], 514). It is, in other words, another disturbing instance of matter's dynamism; of matter, doing what it ought not. But the most extreme example of this Gothicised merging 'of natural and unnatural, ordinary and extraordinary' (Willis 2012 [2014], 18) is the case of Krook, the shopkeeper. The illiterate Krook comes into possession of a Will that might resolve the case of Jarndyce and Jarndyce, but he does not, cannot know its significance. Before Krook can find out the truth, Dickens has him spontaneously combust (Ch. 32). There is an obvious symbolic significance to Krook's death: as an illiterate collector of society's detritus, Krook is surrounded by signs that he cannot read, just as society is itself beset by strange symptoms and warnings whose meaning it cannot divine, but must, if it is to survive a conflagration of its own making. But that is not the only point that Dickens is making: defending his decision to kill off Krook in this most apparently improbable way, Dickens later insisted *that these things happened*; that spontaneous combustion was a matter of recorded fact (Preface; Dickens 2003, 6). What society lacked, in other words, was understanding, and the Gothic marks its limits.

In the Gothicised world of *Bleak House*, therefore, nothing (no thing) behaves as it should, and society is left as baffled and bewildered as it is unnerved. And unnerved it most certainly is: these threatening Gothic entities are everywhere – and nowhere – from the acrid fog (or more properly smog), catching in 'the eyes and throats of ancient Greenwich pensioners' (Ch. 1, 13), to the Court of Chancery; as I discuss in more detail below, Chancery is itself a 'monstrous alienating force' (Willis 2012 [2014], 19) of a scale and with a reach that properly qualifies it as a (Gothic) hyperobject, 'massively distributed in time and space relative to humans' (Morton 2013, 1).

Whilst Dickens is fascinated by what things are, he is also, as John stresses, conscious 'of the constructedness as well as real power of [the] hierarchies of value' that are attached to those things (John 2012, 117). It is not just a question of what things are, but of how they are perceived, and how they are weighed. In Dickens's estimation, no thing is ever nothing, but that is not the nature of a modern society's own workings. It is in the nature of industrial modernity, argued Karl Marx, to create estrangement and alienation: the more energy each worker invested in the product of her or his labour, the more alive it became, but as objects came to seem alive, so people increasingly resembled objects, interchangeable and ultimately disposable (Marx 1975 [1992], 324). Young Jo is a case in point. Jo is 'a crossing-sweeping boy' (Ch. 29, 467), and perhaps the most inconsequential figure in the society depicted by the novel: 'of no order and no place, neither of the beasts nor of humanity' (Ch. 47, 724). Jo presents society with an epistemological conundrum, a mystery that it will harry and demonise. He is, as Parham observes, '[t]he epitome … of "wasted" life, around which the depiction of environmental justice turns' (2017 [2019], 123). Society sees Jo as a mere 'thing', named for the function he performs, as a crossing sweeper, who (for a few pennies) clears the way for the well-to-do through the ordure that covers the streets. Jo as 'waste' is, in other words, all but indistinguishable *from* waste. How has Jo come to this? Lacking a family of his own, orphan Jo has been abandoned – orphaned, as it were – by the state, and by society; their lack of care is all the more egregious because Jo is a little more than a child. At once a 'focus of injustice' (Eagleton 2005, 152) and a 'harrowing [symbol] of oppression' (153), Jo stands as an indictment of the Victorian mantra of 'self-help'; free to work as hard as he can, his only real freedom is to live a short and impoverished life.

Unschooled and ignorant, Jo can make no sense of his situation. 'I don't know nothink [sic]', Jo declares, 'about nothink at all' (Ch. 16, 260). Jo might well seem entirely powerless, but perhaps, the narrator suggests, he is like the oxen on market day, 'over-goaded, over-driven, never guided, run into wrong places'; they 'often sorely hurt the innocent, and often sorely hurt themselves. Very like Jo and his order; very, very like!' (Ch. 16, 258). Or perhaps Jo resembles a 'thoroughly vagabond dog'; 'they are probably upon a par' (Ch. 16, 259). 'Turn that dog's descendants wild, like Jo', adds the narrator, 'and in a very few years they will so degenerate that they will lose even their bark – but not their bite' (Ch. 16, 259). Dickens's brief aside embodies all the limitations of middle-class Victorian commentaries on working-class lives, not least the trap of environmental determinism discussed in Chapter 3, palpably present in the narrator's ready comparison of 'Jo and his order' with domesticated animals, degenerating into wild beasts (Fasick 1996, 139–40) (here, too, we can detect the process of othering that Plumwood describes, as a homogenous 'order' or 'underclass' is aligned with an inferiorised nature). But as the narrative also hints, Jo's mistreatment and neglect is the basis for potential insurrection. There is a suggestion, here, of Engels's own conclusion, that a mistreatment so systematic – an oppression amounting, Engels wrote in *The Condition of the Working Class*, to 'social murder' (1993 [2009], 107) – must and would lead to a mass uprising.[6]

Yet Dickens has another outcome in mind. Jo will indeed become a threat to those with power and prestige, but not as a convert to the cause of revolution. It will simply be because 'Jo lives – that is to say, Jo has not yet died – in

… Tom-all-Alone's' (Ch. 16, 256), and Tom-all-Alone's engenders its own threat to social order:

> It is a black, dilapidated street, avoided by all decent people, where the crazy houses were seized upon, when their decay was far advanced, by some bold vagrants who after establishing their own possession took to letting them out in lodgings. Now, these tumbling tenements contain, by night, a swarm of misery. As on the ruined human wretch vermin parasites appear, so these ruined shelters have bred a crowd of foul existence that crawls in and out of gaps in walls and boards; and coils itself to sleep, in maggot numbers, where the rain drips in; and comes and goes, fetching and carrying fever …
>
> (Ch. 16, 256–7)

As Janice Allan notes, 'fever' was 'a catch-all term for a number of diseases, including typhus, cholera and smallpox' (2003, 33); the very indeterminacy of the term signals the reality of life in a risk society.[7] So does Dickens's curiously disturbing textualisation of the problem that 'fever' represents, in a modernising (but not yet truly modern) society: as his description (above) of the rookery and its inhabitants unfolds, it is no longer clear which (or what) he is describing. Is this 'crowd of foul existence' even human, as it 'coils itself to sleep, in maggot numbers'? Or has it become (as Martin Willis suggests in his own brief reading of the Gothic in *Bleak House*) a 'supernatural animation of disease' (2012 [2014], 21), the literal embodiment of the disease(s) it 'sows in its every footprint' (21)? To describe this shapeless, monstrous 'crowd' as disease – as dis-ease – underlines how unsettling readers would have found Dickens's prose; were they themselves so safe from the spread of this mute yet lethal contagion?

Dickens supplied an answer. Inevitably, Jo falls ill, and just as inevitably, passes on his (undiagnosed and unspecified) illness. In the free-for-all of a competitive, acquisitive, individualistic society, it might seem that Jo exists in isolation, but he does not. As Stacey Alaimo argues, the body is not bounded: it is porous, permeable, and 'radically open' (2010, 13); it becomes a place 'in which social power and material/geographic agencies intra-act' (63). Although Jo is the overlooked product of society's indifference, he is also at the centre of that society, positioned at precisely those intersections of the very poor (who converge on the rookery) and the very rich (whom Jo will seek out for the penny or two he might get from them). As the narrator later observes, 'how strangely fate has entangled this rough outcast in the web of very different lives' (Ch. 47, 732).

Thus, Jo's story echoes the one that Carlyle tells in *Past and Present* (1843), of the 'poor Irish widow' who, infected with 'typhus-fever', is refused all help, only to infect and kill 17 of 'her fellow-creatures' before herself dying (Carlyle 1843 [1897], Vol 6, 359): was this not demonstration enough, thundered Carlyle, 'that she *was* flesh of your flesh'? (italics in the original; 360). In *Bleak House*, the Irish widow is Jo, representative of the residuum that society has disowned: he is society's victim, mute and uncomprehending, but now himself agential, with the shocking potential to undo the lives of all those with whom he comes into contact; he has himself become the conduit for Dickens's Gothicised 'fever', 'given consciousness enough to sow evil' (Willis 2012 [2014], 21). What then comes of the threat that Jo poses to society? Does the disease he carries perhaps rebound on the rich and the privileged, 'breaking up the pattern of class ... society', as Beck predicted (1992 [2012], 23)? In a cruel twist, Jo infects Esther, one of the few people to have ever taken a genuine interest in him, and herself another of those figures whom society disvalues. Jo's fault is to be poor; Esther's is to be a woman, and worse, one born out of wedlock. Yet it falls to Esther to step in and do what a patriarchal society does not, and care for Jo.

Despite her best efforts, however, Jo will not long survive. Who or what, then, is responsible for Jo's death? As Dickens emphasised, Jo's fate is entangled with the existence of Tom-all-Alone's, the novel's most obvious and egregious instance of environmental injustice: a rookery that 'propagates infection and contagion' (Ch. 46, 710), it is still the only place where Jo can live on the few pennies he manages to make. Here, those with very little exploit those with still less, in a grotesque working out of the logic of political economy (Beames, above; 1852, 2). But the origins of Tom-all-Alone's lie elsewhere, in the machinations of another material-discursive structure. As Dickens wastes no time explaining, '[t]his desirable property is in Chancery' (Ch. 16, 257). It is, in other words, caught up (and its origins lost) in a court case; 'the suit had laid the street waste' (Ch. 16, 257); and responsibility should be laid before 'the Divan of justice' that is the Court of Chancery (Ch. 47, 720).

'[M]istily engaged' in advancing – or retarding – any number of the causes it has entrapped, 'and making a pretence of equity' (Ch. 1, 14), Chancery's obscure machinations draw together the (apparently unconnected) characters in the novel, and blight many of their lives; it is at the heart of a novel which stands as an indictment of 'all the injustice it has committed and all the misery it has caused' (Ch. 1, 19). Chancery was founded, in part, to provide a discretionary alternative to the common law: every case would be judged on its merits, providing a system that was in theory equitable and fair (Allan 2003, 18). Some still saw Chancery 'as the purest fount of Justice', as Alfred Whaley Cole observed in 'The Martyrs of Chancery', an article published in Dickens's *Household Words* in 1850 (251). By the middle of the nineteenth century, as Cole pointed out and Dickens had a personal reason to know, the Court of Chancery was corrupt and inefficient (Allan 2003, 19). It made prisoners, wrote Cole (1850, 250), of those who were entangled in it, and stranded them in a kind of helpless limbo of indefinite duration, as Tom-all-Alone's was itself stranded, 'cut off from honest company and put out of the pale of hope' (Ch. 16, 257).

What has happened, then, to create from a court that was intended to provide a better quality of justice a kind of machine for the creation of its opposite? By what malign spirit has it, in turn, been infected? Once again, it might be said that money is at the root of it, since Chancery paid for itself from the trusts, estates, and legacies whose fate it decided, establishing the basis for a self-perpetuating system whose rituals and obfuscations were entirely self-serving. But the nature of those rituals and obfuscations creates a further problem, as questions of responsibility become ever more attenuated. Just as no one now knows whether the eponymous 'Tom' of Tom-all-Alone's was the plaintiff or the defendant in the case, or in any way connected to it (Ch. 16, 257), no one knows who, exactly, is running the Court of Chancery. It has become, quite simply, a kind of all-consuming monster, indifferent and autonomous, a Gothicised hyperobject that generates its own Kafkaesque nightmare of lost souls. It cannot even be said to serve the privileged, giving 'to monied might the means of abundantly of wearying out the right' (Ch. 1, 15): as that 'Monument of Chancery practice' (Ch. 65, 975), the case of Jarndyce and Jarndyce, will eventually demonstrate, it is capable of consuming a fortune of any size, thereby demonstrating the same commendably democratic impulse as the fevers that Tom-all-Alone's engenders, working their way through 'every order of society up to the proudest of the proud and to the highest of the high' (Ch. 46, 710).

That is the price of entanglement – of entangled lives, entangled entities – when the responsibilities that entanglement implies are forgotten. Dickens sides with Esther: even as no one claims or admits responsibility, Esther insists that we all are responsible (Eagleton 2003, x). Nor is Jo the only victim of this indifference, this isolationism. As the narrator declaims, on the occasion of Jo's death:

> Dead, your Majesty. Dead, my lords and gentlemen. Dead, right reverends and wrong reverends of every order. Dead, men and women, born with heavenly compassion in your hearts. And dying thus around us every day.
>
> (Ch. 47, 734)

This is, Nicola Bradbury suggests, 'Dickens's most celebrated rhetorical out burst', and he comes 'out strongly, unequivocally, on behalf of the oppressed and the silent' (2003, xxxiii).

The question, nevertheless, is how to address this continuing tragedy, and shape a society that might in the future avoid the obvious injustices of a Tom-all-Alone's. Here, Dickens hesitates. If we are all implicated in the lives of others, as Esther believes, we are all equally responsible for them; but Esther's best efforts cannot save poor Jo. Even Woodcourt, the doctor whom Esther will eventually marry, can do nothing for him. Like many of the threats that industrial modernity engenders, the unnamed 'fever' that carries off poor Jo lies out of reach of human understanding and remedy. Meanwhile, at

Tom-all-Alone's, 'the people "have been down by dozens" and are carried out dead and dying "like sheep with the rot"' (Ch. 22, 358). Viruses might well be democratic, but the society upon which they visit themselves still loads the odds in favour of the rich.[8] *This* lies beyond Esther's – or Dickens's – power to fix. The problem, Eagleton has argued, is that Dickens distrusts and fears 'the kinds of social forces which might [help] to transform nineteenth-century England into the place of common responsibility he pleads for so eloquently' (2005, 162). Not for Dickens any revolutionary alternative.[9] He is similarly doubtful of all forms of system or institution (Fasick 1996, 137), a scepticism which extends to the state itself: what would it become, if it took on the kind of collective responsibility for which *Bleak House* pleads? Would it not simply turn into one vast Court of Chancery or Circumlocution Office, which no one understands and no one can control, and where, absolved of responsibility, no one cares to try? Or would its efforts to care for the disadvantaged not culminate, once again, in oppressive systems like the Poor Law, and in institutions like the workhouse? Not the least most challenging aspect of Dickens's scepticism is that it anticipates a growing view within environmental justice studies itself, that, contrary to the 'classic progressive-Left' argument, states and state institutions are themselves *inherently* predisposed to perpetuate both their own power and 'various forms of inequality (including institutional racism and speciesism)' (Pellow 2018, 23).

Finding no answers to these questions in the novel, the reader is left with the challenge posed by Esther's belief that, 'if nobody is responsible ... we all must be' (Eagleton 2003, x): that we must all step in to address inequality and injustice. As Eagleton argues, 'this belief that we all have a stake in each other' (2003, ix–x) is its own form of resistance to 'the dominant Victorian philosophy of selfish individualism' (x). Yet this apparently simple injunction – an injunction which returns the focus to the individual, as opposed to the collective in the form of state or society in the abstract – is itself profoundly problematic, for reasons that *Bleak House* illustrates. Confronted by the possibility that we are all entangled not only with each other but with all those 'Others' (such as nonhuman animals) that dualism discounts, we are also forced to confront 'a new sense of a-count-ability, a new arithmetic, a new calculus' (Barad 2010, 251). As Barad has argued:

> Responsibility is not an obligation that the subject chooses but rather an incarnate relation that precedes the intentionality of consciousness. Responsibility is not a calculation to be performed. It is a relation always already integral to the world's ongoing intra-active becoming and not-becoming. It is an iterative (re)opening up to, an enabling of responsiveness.
>
> (2010, 265)

Yet this 'enabling of responsiveness' assumes what Dickens himself seemed, increasingly, to doubt, as his novels grew darker and their tone markedly less optimistic. In a society in which Christian compassion takes second place to self-interest, what guarantee is there that anyone will be minded to take up the challenge? Not everyone is an Esther. Some are like Sir Leicester Dedlock, his feelings fossilised by long custom, who is concerned only to preserve his own family name. Some are like Snagsby, who drops a half-crown into the hand of poor Jo ('his usual panacea for an immense variety of afflictions' [Ch. 22, 362]), without ever wondering whether they are more like him, or where they might live when they are not sweeping the street corner. Some are like Harold Skimpole, who claims that he is 'a mere child in the world' (Ch. 6, 90), and blithely evades all his responsibilities as friend, husband, and father.

The still more knotty problem is that, whilst Esther may well believe that we are all equally responsible for those whom society has disadvantaged (Eagleton 2005, 161), she neglects to mention what the narrative has made transparently clear: that not everyone *is* equally responsible, and to claim as much risks allowing the culpable and the blameworthy to escape into 'a soup of post- (or pre-) modern complexity' (Haraway 2008, 88). Furthermore, what is true of responsibility is also true of what Donna Haraway calls 'response-ability' (2008, 89): just as we are not all equally responsible, so we are not all equally able to respond. Esther herself is amongst the least responsible for Jo's fate, *and* the least able to do anything about it; she is simply an orphaned young woman in a society that systematically strips women of public power and authority, and denies that what they do as care-givers and house-keepers *is* work (Danahay 1991, 419).[10] That powerlessness is, after all, reflected in her own, tortured first-person narrative, a narrative in which she contrives to offer up her own point of view whilst simultaneously disavowing her ability and right to do so. By contrast, those who bear most responsibility for society's injustices – and those who have the greatest power to address them – are also those who make no effort to do so. The Lord High Chancellor is a case in point: wrapped in his 'foggy glory' (Ch. 1, 14), he is apparently oblivious (or indifferent) to the faults and failings of a system whose very purpose is to deal in justice, and provide it to those who feel that they had otherwise been failed by the law. Instead, he presides over an institution that makes possible the ruinousness of Tom-all-Alone's, and the human misery it engenders.

In one of Dickens's earlier works, the Lord High Chancellor might, perhaps, have been the villain of the piece, finished off in the final act by the lightning bolt of Dickens's righteous indignation. Yet *Bleak House* is of a different order to a novel like *Oliver Twist*, with its Fagin, destined for the hangman's noose. Here as in Dickens's other later novels, society's problems resist that kind of easy solution, that straightforward (but in its own way profoundly problematic) catharsis. Like the lawyers by whom he is surrounded, the Lord High Chancellor is allowed to evade responsibility, vanishing, as it were, into the London fog that stands so effectively not just for Chancery's obfuscation, but for the complexities of a modern society that defies comprehension. It is a seemingly bleak conclusion. As Michael

Slater observes, Dickens presents himself as the romantic visionary (2009, 10), who can part modernity's sea of confusion, and reveal the connexions that London's frenetic dynamism obscures, thereby demonstrating that not only is this alien, urban monster knowable but itself susceptible to justice. *Bleak House* is, as Eagleton observes, 'a forensic kind of fiction' (2003, ix). Yet the reader is left with the impression of unfinished business, of justice incomplete; of 'a tissue of loose ends and questions rather than of neatly resolved patterns' (Miller 1991, 197), and of an already post-modern world in which meaning is constantly deferred (Lesiuk 2018, 124). What is to be done, therefore, when instances of environmental injustice still humble society's claim to be modern, progressive, civilised, and when the nature of that society may itself conspire against a resolution? Perhaps this is a reason why, as Bruno Latour has argued (1993), society has not yet succeeded (and may never succeed) in becoming truly modern: '[i]n the countless quarrels between Ancients and Moderns', Latour observed,

> the former come out winners as often as the latter now, and nothing allows us to say whether revolutions finish off the old regimes or bring them to fruition. Hence the scepticism that is oddly called 'post' modern even though it does not know whether or not it is capable of taking over from the Moderns.
>
> (1993, 10)

In the midst of this confusion, Dickens insists, there remains a moral imperative to act, an imperative which, in a characteristically quiet way, Esther establishes as her aim near the start of her narrative: 'to render what kind services I could, to those immediately about me; and to try to let that circle of duty gradually, and naturally expand itself' (Ch. 8, 128). As Parham remarks, Esther's 'character represents … the necessity of acting, however hesitant and contingent those actions might be, towards developing material *and* social ties that nurture rather than wither human and nonhuman life' (2017 [2019], 127). She embodies, in other words, what might otherwise seem impossible: the possibility of an emergent justice, a justice-to-come (Derrida 1992, 26–7). As James Martel observes, 'justice *must* be immediate, but it remains aloof, just out of reach, and (therefore) requires our own response in the process' (2011, 161). It requires, as Esther insists, our own, direct involvement, our own constant negotiation and renegotiation.

The odds of advancing that kind of just and fair alternate are, all too obviously, stacked against Esther, who is (or by rights should be) a marginal figure, undermined by a sense of her own inadequacy, haunted by the thought of her own illegitimacy, and blighted by illness. Yet she succeeds not only in surviving, but in thriving. The novel closes with Esther's own account of the community she has contrived to create around the second Bleak House, 'a rustic cottage of doll's rooms' (Ch. 64, 962) whose communal warmth (like that of the first Bleak House) belies its inauspicious, gloomy name. '[I]t is right to begin with the obligations of home', she tells Jarndyce (Ch. 6, 83) – 'a good man, superior to injustice' (Ch. 24, 403) – but not because she is simply mimicking society's insistence that, if she is to be a respectable, well-to-do woman, this is and must be her role. It is rather because she has seen society for what it is, as a mere 'agglomeration of practical people' (Ch. 6, 91), more concerned with making money than in honouring each other, and she has seen that it will not do. 'We are not rich in the bank', writes Esther, as she looks back over her married life, 'but we have always prospered' (Ch. 67, 988). She might equally have said: we have always flourished.

*

In *Bleak House*, Dickens struggles with the complex reality of a world reshaped by industrial modernity, a world in which environmental injustice is an urgent and palpable problem yet responsibility for it is somehow evaded – perhaps unwittingly, perhaps wilfully. His answer lies in a novel whose vast scale and apparently chaotic mix of narratives and characters slowly yields to meaning. Yet the reality of entanglement, as itself an iterative process that constantly defers resolution, conspires against any kind of simplistic understanding of, or totalising solution to it. Dickens's answer is a modest one, but in its own way radical; he leaves the future to Esther, a self-effacing character in a society that has itself consistently ignored her (as a woman, as an orphan, as illegitimate, as a care-giver). Esther does not claim to have the answers; she insists only on the need to take responsibility, in the world that lies immediately to hand; to reach out through the relationships she builds to a better, fairer future. That future lies outside and beyond all those institutions and entities which, like the hyperobjects that Morton describes (2013), defy human comprehension and control; it lies in the haven she creates in the midst of a self-interested and turbulent society. Thus, and whilst Esther has failed to address all of society's ills and injustices, she has nevertheless found a way to refashion that society on a scale that might itself permit their resolution, in so doing achieving a least a measure of success 'in opposing the chaos of a world tainted by injustice' (McClure 2003, 40).

Significantly, Esther's haven takes as its setting a benign, green world, 'tranquil and so beautiful' (Ch. 64, 962), in which human and nonhuman coexist in a necessary partnership. These engagements with the natural world constitute a dimension of human experience – a constitutive dimension – that industrial modernity was quickly and systematically erasing, and that erasure was nowhere more obvious than in the factory towns that sprang up as if unbidden in the early years of the Victorian period. It was to just this kind of abnatural environment that Dickens turned in *Hard Times* (1854), published immediately after *Bleak House*. As I discuss in the next chapter, *Hard Times* is at once an extended and critical response to the kind of environmental classism that Engels had uncovered in Manchester, and a continuing dialogue about the nature of a flourishing existence – and the importance of the natural world to it.

*

Notes

1 This is just one reading of – one potential approach to – a novel that has (as I noted in the Introduction) become something of an ur-text for ecocritics. In *Victorian Literature, Energy, and the Ecological Imagination* (2014), for example, Allen MacDuffie discusses 'energy consumption as a pressing environmental issue' (2014, 15), and includes a chapter on *Bleak House* (2014, 89–113): Dickens' 'great subject', MacDuffie writes, is 'life in the world's first energy-intensive economy' (115), its impacts and effects. In his own discussion of *Bleak House* (21–43), Jesse Oak Taylor analyses the 'manufactured climate' (2016, 21) that is the most obvious consequence of that economy. Leading British ecocritic John Parham also discusses *Bleak House* and the 'human material ecology' it describes (2017 [2019], 114), a discussion which relates environmental injustice to society's dismissal of matter (as 'waste') and those marginalised groups within society that it also exploits (116); Parham's insights are discussed in more detail elsewhere in this chapter. Michael Lesiuk (2018) has also related Dickens' use of metonymy in the novel to Timothy Morton's concept of the mesh (in Morton's *The Ecological Thought*, 2010), which can be read across into my own emphasis on the novel's presentation of an entangled and agential reality, whose shifting meanings inevitably unsettle any attempt at understanding: it is the nature of that reality which calls for (and brings forth) the figure of the inspector, Bucket, 'man in the abstract' (Dickens 2003, Ch. 53, 803). For a feminist ecocritical/material feminist approach to *Bleak House*, and a further discussion of Esther's role in the novel, the reader may also wish to seek out my own approach to '*Bleak House*, Esther Summerson, and Gendered Identity' in *Dickens & Women ReObserved* (2020).

2 The Penguin edition (2003) of *Bleak House* is the copy text for this chapter, and hereafter, all references take the simplified format of chapter and page number.

3 In his depiction of Tom-all-Alone's, notes Allan Pritchard, Dickens transposes the traditional Gothic setting – the remote and gloomy country house or castle (1991, 435) – and re-establishes it in an urban context: 'the Gothic ruin ... is now represented by the slums and dilapidation of the great city' (437). In a complementary reversal, it will transpire that the Bleak House of the novel's title is (somewhat ironically) a place of familial warmth and comfort, notwithstanding its windswept rural setting and tragic history.

4 London's smogs would become so famous – and so definitive an example of the phenomenon – that scientists continue to use the term 'London type' smog to distinguish this 'classical' form of it from photochemical smog (itself named after a city – Los Angeles) (Perez-Diaz et al. 2017, 11).

5 As I discussed in the Introduction, this has ethical as well as epistemological consequences: if we cannot distinguish ourselves from the natural or nonhuman world western societies so routinely disregard and discount as (economic) externality – if, in other words, we accept this ontological flattening – we must also acknowledge that intrinsic worth is not confined to humans (still less some humans rather than others), but everywhere present; it is on this basis that we can redefine the terms of environmental justice itself, as an emergent justice that (albeit problem-

atically) takes into account the nonhuman, the natural, the material world in all its own aliveness (Barad 2007, 33). But the starting point for this radical reworking of what constitutes justice – and to what or whom it should apply – is the blurring of the hitherto rigid categories of subject and object.

6 That hint can also be traced back to the term 'rookery' itself, since the collection noun for rooks is a 'parliament', signifying the widespread belief that rooks possessed 'an innate democratic spirit' (Cocker 2007 [2016], 125), 'held moral views of their fellows' conduct' (121), and would, on occasion, assemble to judge and punish wrongdoing (122).

7 Medicine was itself only slowly becoming scientised; thus far, it had proven to itself only its ineffectuality, giving rise to a strong vein of medical nihilism that with good reason persisted throughout the century (Stegenga 2018, 11). Thus, Jo succumbs, simply, to 'fever' (Ch. 31), and it is an unnamed 'fever' that he passes on to Esther. Whatever the illness, Dickens does not name it, any more than he names any of the other illnesses in the novel (Carpenter 2010, 92). Nor does he need to: this vagueness, this uncertainty is itself part of the reality he sought to depict. Yet Dickens does attribute to Esther symptoms specific enough for medically minded readers to identify the illness as smallpox, for which, quite remarkably, a vaccine had been available for over half a century; a Compulsory Vaccination Act was passed in the same year that *Bleak House* completed its run at the publishers (1853) (Carpenter 2010, 92). The fact that no one has thought to vaccinate Esther – and Esther has herself failed to demand it – is, perhaps, further proof of her marginal status, and the process of introjection through which she has internalised that marginalisation (see also Tait 2020, 126–8).

8 Industrial modernity did not produce the 'fevers' to which Dickens refers, but the conditions it generated made their spread more likely and their impact more dramatic amongst those whom it disadvantaged. Cholera is a case in point. An Asiatic arrival, it was carried to Britain along the nation's now globalised trade-routes, and first struck in 1831. Within the year, cholera had killed over 20,000 in the first of four major epidemics. In 1854, an outbreak in Soho caused 'upwards of five hundred fatal attacks of cholera in ten days', a mortality rate that rivalled the plague (Snow 1855, 38). But cholera did not affect everyone, equally. The Soho outbreak is an example. Although Soho adjoined the affluent City of Westminster, it was not a wealthy neighbourhood, and the clue to the socioeconomic status of the outbreak's victims can be found in the way in which they contracted the disease. Famously, Dr. John Snow traced the outbreak to a single hand pump (Snow 1855, 38–41; see also Steinbach 2017, 16), perforce shared by all the local families, none of whom could afford to pay for the private supply of fresh water. In *Bleak House*, Dickens makes the same point, but in a more complex way, which reflects the instantiation of new risk positions *within* a risk society; with the way, in other words, that even when threats appear pervasive and unavoidable – and distribution of them 'democratic' – a greater burden of them may nevertheless fall on the disadvantaged. Esther catches her unnamed illness even though she is relatively privileged, and contracts her illness whilst living in rural Bleak House. Neither relative affluence nor a life in the country are sufficient protection, as the mid-century spread of scarlet fever underlined (Davenport 2020, 480). There are, moreover, few cures for sickness. Yet the fact remains that the well-fed and well-cared-for have a much better chance of survival than the cold, hungry, and destitute. Jo dies; Esther lives; and in the slum from which Jo hales, the dead are carried out 'by dozens' (Ch. 22, 358).

9 As his portrayal of the workers or 'hands' in *Hard Times* suggests, Dickens felt threatened by working class attempts to assert their own collective identity and insist on their own independence. The historical irony is that Dickens was writ-

ing in the aftermath of the Year of Revolutions, 1848, which was in the British Isles notable for the absence of a revolution. Whether or not Chartism ever constituted a revolutionary threat – a matter for debate – the movement had given way to a spirit of cooperation and conciliation embodied in the rise of respectable trade unions. It was only several decades later that the working-class movement once again took up a more assertive, aggressive stance, and only then because a hegemonic middle class had failed to address the iniquitous living and working conditions to which Dickens himself had repeatedly drawn its attention.

10 Work such Esther's – her work as a home-maker and care-giver – is essential to a patriarchal, heteronormative society intent on perpetuating itself and its own values, yet that work is (as Plumwood suggests) denied, ignored, and pushed into the background (1993 [1997], 48); to admit its importance would be to admit (in part) the faults and failings of the dominant structures of meaning, which assign value to the master but not to 'the inferiorised other' (48).

Works cited

Alaimo, Stacey. 2010. *Bodily Natures: Science, Environment, and the Material Self*. Bloomington, IA: Indiana University Press.

Allan, Janice, ed. 2003. *Charles Dickens's Bleak House: A Sourcebook*. London: Routledge.

Barad, Karen. 2007. *Meeting the Universe Halfway: Quantum Physics and the Entanglement of Matter and Meaning*. London: Duke University Press.

Barad, Karen. 2010. 'Quantum Entanglements and Hauntological Relations of Inheritance: Dis/continuities, SpaceTime Enfoldings, and Justice-to-Come'. *Derrida Today* 3 (2): 240–68. doi: 10.3366/E1754850010000813

Beames, Thomas. 1852. *The Rookeries of London: Past, Present and Prospective*. 2nd edn. London: Thomas Bosworth.

Beck, Ulrich. 1992 [2012]. *Risk Society: Towards a New Modernity*, translated by Mark Ritter. London: Sage.

Bennett, Jane. 2010. *Vibrant Matter: A Political Ecology of Things*. London: Duke University Press.

Bradbury, Nicola. 2003. 'Introduction'. Charles Dickens, *Bleak House*, edited and introduced by Nicola Bradbury, xix–xxxv. London: Penguin.

Briggs, Asa. 1963 [1987]. *Victorian Cities*. Harmondsworth: Penguin.

Briggs, Asa. 2003. *Victorian Things*. Rev. edn. Stroud: Sutton Publishing.

Brimblecombe, Peter. 1987. *The Big Smoke: A History of Air Pollution in London Since Medieval Times*. London: Methuen.

Buell, Lawrence. 2001. *Writing for an Endangered World: Literature, Culture, and Environment in the U.S. and Beyond*. London: Belknap Press.

Carlyle, Thomas. 1843 [1897]. *Past and Present. The Works of Thomas Carlyle, Volume Six*, 417–500. New York, NY: Peter Fenelon Collier. Available at archive .org, https://archive.org/details/worksofthomascar0006carl/page/n5/mode/2up ?view=theater. Accessed 15 November 2022.

Carpenter, Mary Wilson. 2010. *Health, Medicine, and Society in Victorian England*. Santa Barbara, CA: Praeger.

Chinn, Carl. 1995 [2006]. *Poverty Amidst Prosperity: The Urban Poor in England, 1834–1914*. Lancaster: Carnegie.

Cocker Mark., 2007 [2016]. *Crow Country: A Meditation on Birds, Landscape and Nature*. London: Vintage.

Cole, Alfred Whaley. 1850. 'The Martyrs of Chancery'. *Household Words*, 7 December: 250–2. Available at archive.org, https://archive.org/details/householdwords02dicklond/page/250/mode/2up?q=martyrs. Accessed 10 February 2021.

Coolsaet, Brendan, and Pierre-Yves Néron. 2021. 'Recognition and Environmental Justice'. *Environmental Justice: Key Issues*, edited by Brendan Coolsaet, 52–63. London: Routledge.

Danahay, Martin A. 1991. 'Housekeeping and Hegemony in *Bleak House*'. *Studies in the Novel* 23 (4): 416–31. Available at jstor.org, http://www.jstor.org/stable/29532815. Accessed 7 December 2022.

Davenport, Romola J. 2020. 'Urbanisation and Mortality in Britain c.1800–1850'. *Economic History Review* 73 (2): 445–85. doi: org/10.1111/ehr.12964.

Derrida, Jacques. 1992. 'Force of Law: 'The Mystical Foundation of Authority''. *Deconstruction and the Possibility of Justice*, edited by Drucilla Cornell, Michel Rosenfeld, and David Gray Carlson, 3–67. Abingdon: Routledge.

Dickens, Charles. 1854 [1998]. 'To Working Men'. Michael Slater (ed.). *'Gone Astray' and Other Papers from Household Words*, 1851–1859, 225–9. London: J. M. Dent.

Dickens, Charles. 2002. *Dombey and Son*, edited and introduced by Andrew Sanders. London: Penguin. First published 1848.

Dickens, Charles. 2003. *Bleak House*, edited and introduced by Nicola Bradbury, preface by Terry Eagleton. London: Penguin.

Dyos, H. J. 1967. 'The Slums of Victorian London'. *Victorian Studies* 11 (1): 5–40. Available at jstor.org, www.jstor.org/stable/3825891. Accessed 27 January 2022.

Eagleton, Terry. 2003. 'Preface'. Charles Dickens, *Bleak House*, edited and introduced by

Nicola Bradbury, preface by Terry Eagleton, vii–xii. London: Penguin.

Eagleton, Terry. 2005. *The English Novel: An Introduction*. Oxford: Blackwell.

Engels, Friedrich. 1993 [2009]. *The Condition of the Working Class in England*, edited by David McLellan. Oxford: Oxford University Press.

Fasick, Laura. 1996. 'Dickens and the Diseased Body in *Bleak House*'. *Dickens Studies Annual* 24: 135–51. Available at jstor.org, www.jstor.org/stable/44372460. Accessed 7 December 2022.

Forster, John. Not dated. *The Life of Charles Dickens*. London: Chapman & Hall. First published 1872–4.

Garwood, John. 1853. *The Million-Peopled City; or, One-Half of the People of London Made Known to the other half*. London: Wertheim and Macintosh.

Hansard HC [House of Commons] Debate 129 (1495–7), 8 August 1853. Available from hansard.parliament.uk, https://hansard.parliament.uk/Commons/1853-08-08/debates/a4437a20-326e-4b19-a4ab-232fe434221e/SmokeNuisanceAbatement(Metropolis)Bill. Accessed 27 October 2021.

Haraway, Donna J. 2008. *When Species Meet*. London: University of Minnesota.

Hilton, Boyd. 2006. *A Mad, Bad, and Dangerous People? England 1783–1846*. Oxford: Clarendon Press.

Himmelfarb, Gertrude. 1984. *The Idea of Poverty: England in the Early Industrial Age*. New York, NY: Alfred A. Knopf.

Hoppen, K. Theodore. 1998. *The Mid-Victorian Generation*: 1846–1886. Oxford: Clarendon Press.

John, Juliet. 2012. 'Things, Words, and the Meanings of Art'. *Dickens and Modernity*, edited by Juliet John, 115–32. Cambridge: D. S. Brewer.

Joyce, Simon. 2002. 'Inspector Bucket versus Tom-all-Alone's: *Bleak House*, Literary Theory, and the Condition-of-England in the 1850s'. *Dickens Studies Annual* 32: 129–49. Available at jstor.org, www.jstor.org/stable/44372054. Accessed 7 December 2022.

Latour, Bruno. 1993. *We Have Never Been Modern*, translated by Catherine Porter. Cambridge, MA: Harvard University Press.

Lesiuk, Michael. 2018. 'Reading the Mesh of Metonymy in *Bleak House*'. *The Journal of the Midwest Modern Language Association* 51 (1): 121–45. Available at jstor .org, www.jstor.org/stable/45017408. Accessed 7 December 2022.

Levine, Caroline. 2012 [2013]. 'Victorian Realism'. *The Cambridge Companion to the Victorian Novel*, edited by Deidre David, 84–106. Cambridge: Cambridge University Press.

Lougy, Robert E. 2002. 'Filth, Liminality, and Abjection in Charles Dickens's *Bleak House*'. *ELH* 69 (2): 473–500. Available at jstor.org, www.jstor.org/stable /30032028. Accessed 7 December 2022.

MacDuffie, Allen. 2014. *Victorian Literature, Energy, and the Ecological Imagination*. Cambridge: Cambridge University Press.

Martel, James. 2011. 'Waiting for Justice: Benjamin and Derrida on Sovereignty and Immanence'. *Republics of Letters: A Journal for the Study of Knowledge, Politics, and the Arts* 2 (2): 158–72. Available at https://arcade.stanford.edu/sites/default /files/article_pdfs/roflv02i02_Martel_060111_0.pdf. Accessed 12 January 2023.

Marx, Karl. 1975 [1992]. 'Economic and Philosophical Manuscripts (1844)'. *Early Writings*, translated by Rodney Livingstone and Gregor Benton, 279–400. London: Penguin.

McClure, Joyce Kloc. 2003. 'Seeing through the Fog: Love and Injustice in *Bleak House*'. *The Journal of Religious Ethics* 31 (1): 23–44. Available at jstor.org, www.jstor.org/stable/40018201. Accessed 7 December 2022.

Miller, J. Hillis. 1991. 'Interpretation in Bleak House'. *Victorian Subjects*, 179–99. Durham, NC: Duke University Press. Originally published in 1971.

Morton, Timothy. 2010. *The Ecological Thought*. London: Harvard University Press.

Morton, Timothy. 2013. *Hyperobjects: Philosophy and Ecology after the End of the World*. London: University of Minnesota Press.

Mumford, Lewis. 1961 [1973]. *The City in History: Its Origins, its Transformations, and its Prospects*. Harmondsworth: Penguin.

Parham John., 2017 [2019]. 'Bleak Intra-Actions: Dickens, Turbulence, Material Ecology'. *Victorian Writers and the Environment: Ecocritical Perspectives*, edited by Laurence W. Mazzeno and Ronald D. Morrison, 114–29. Abingdon: Routledge.

Pellow, David Naguib. 2018. *What is Critical Environmental Justice?* Cambridge: Polity.

Perez-Diaz, Jose Luis, Ognyan Ivanov, Zahary Peshev, Marco A. Álvarez-Valenzuela, Ignacio Valiente-Blanco, Tsvetina Evgenieva, Tanja Dreischuh, Orlin Gueorguiev, Peter V. Todorov, and Ashok Vaseashta. 2017. 'Fogs: Physical Basis, Characteristic Properties, and Impacts on the Environment and Human Health'. *Water* 9 (807): 1–21. doi: org/10.3390/w9100807

Plumwood, Val. 1993 [1997]. *Feminism and the Mastery of Nature*. London: Routledge.

Pritchard, Allan. 1991. 'The Urban Gothic of *Bleak House*'. *Nineteenth-Century Literature* 45 (4): 432–52. Available at jstor.org, https://doi.org/10.2307/3044996. Accessed 6 December 2022.

Schlosberg, David. 2007. *Defining Environmental Justice: Theories, Movements, and Nature*. Oxford: Oxford University Press.

Slater, Michael. 2009. *Charles Dickens*. London: Yale University Press.

Snow, John. 1855. *On the Mode of Communication of Cholera*. 2nd edn. London: John Churchill. Available at archive.org, https://archive.org/details/b28985266/page/n3/mode/2up?q=&view=theater. Accessed 10 November 2021.

Stegenga, Jacob. 2018. *Medical Nihilism*. Oxford: Oxford University Press.

Steinbach, Susie L. 2017. *Understanding the Victorians: Politics, Culture and Society in Nineteenth-Century Britain*. 2nd edn. Abingdon: Routledge.

Tait, Adrian. 2020. 'Bleak House, Esther Summerson, and Gendered Identity: A Material Feminist Rereading'. *Dickens & Women ReObserved*, edited by Edward Guilliano, 119–35. Brighton: Edward Everett Root.

Taylor, Jesse Oak. 2016. *The Sky of Our Manufacture: The London Fog in British Fiction from Dickens to Woolf*. London: University of Virginia Press.

Williams, Raymond. 1973 [1975]. *The Country and the City*. New York, NY: Oxford University Press.

Willis, Martin. 2012 [2014]. 'Victorian Realism and the Gothic'. *The Victorian Gothic*, edited by Andrew Smith and William Hughes, 15–28. Edinburgh: Edinburgh University Press.

5 Beyond class, gender, species? Charles Dickens's *Hard Times*

In *Hard Times*, first published in 1854, Charles Dickens turned his attention to the factory town, a newly emergent product of the Industrial Revolution, and a locus of what we would now characterise as environmental classism: the systematic environmental impoverishment of an entire class (Bell 2020, 3).[1] Whereas Dickens's previous novel *Bleak House* (1852–3) represents a kind of environmental justice mystery, in which the relationship between the injustice of London's slums and the nature of modern society is not immediately apparent – and responsibility is all too easily evaded – *Hard Times* makes it quite clear what is responsible for the problem it confronts: the convergence of capital, industry, and modernity's discourse of disenchantment, a convergence of structural forces personified in the characters of Josiah Bounderby and his 'eminently practical friend' (Bk. I, Ch. 3, 15), Thomas Gradgrind.[2] It is they who have fashioned a defiantly unequal, polluted, and toxic world. It is also, however, a world that, as Ulrich Beck suggests (1992 [2012], 23), may rebound in unexpected ways on those who have created and most benefit from it.

Unlike the injustice of the slum quarter (or 'rookery') described in *Bleak House*, the iniquities with which *Hard Times* contends are simultaneously self-evident *and* insidious. At its most obvious, environmental classism takes the form of a spatially segregated society, in which the working poor are crowded into heavily polluted slums near the factories where they work, at risk to themselves but not their employers. This is the world that Dickens sketches so memorably in his opening depiction of 'Coketown', the 'keynote' of the novel (Bk. I, Ch. 5, 26), and the embodiment of what ecocritic Elizabeth Carolyn Miller describes as 'an extraction-based industrial society' (2021, 2).[3] As historian John Stevenson observes,

> [f]ew paragraphs have done more … to envision the social repercussions of industrialization: the creation of mass urban society, governed by the regime of the factory and the pace of the machine, an environment polluted and despoiled, and its inhabitants rendered anonymous and dehumanised.
>
> (1993, 229)

DOI: 10.4324/9780367821609-6

But as Stevenson's comments suggest, there is more to environmental classism than the obvious problems of smoky skies, polluted waterways, crowded slums, or the risks of the workplace. Under the conditions created by industrial modernity, impoverishment and injustice take on subtle new forms – including forms of violence which are symbolic and 'slow' – that distort the lives of even those who appear outwardly healthy. In *Hard Times*, Coketown's workers are crippled by a particular way of seeing the world, a discursive legacy of Enlightenment rationality that Dickens sums up with the word 'Fact': 'Fact, fact, fact, everywhere in the material aspect of the town; fact, fact, fact, everywhere in the immaterial' (Bk. I, Ch. 5, 27). In Dickens's novel, 'Fact' stands for 'reason and calculation' (Bk. I, Ch. 15, 93), 'gauging fathomless deeps' with a 'little mean excise-rod' (Bk. III, Ch. 1, 206). In Dickens's estimation, to measure is also to master, to reason, to regulate and control, in so doing 'annihilating the flowers of existence' (Bk. III, Ch. 1, 206); it is to set 'Fact' against 'Fancy', a crucial term in Dickens's critique. As Theodor W. Adorno and Max Horkheimer would later insist, '[t]he program of the Enlightenment was the disenchantment of the world; the dissolution of myths and the substitution of knowledge for fancy' (1997 [2010], 3).

It is this reductive, rationalistic view of the world – an Enlightenment inheritance that insists on the primacy of reason – that underpins the monotony and discipline of a 'severely workful' environment (Bk. I, Ch. 5, 25). The squalor of the slums may be the accidental by-product of the industrialising imperative, readily overlooked by the well-to-do as they moved to the suburbs or beyond, but Coketown's joyless environment is a deliberative and controlling construct, put in place to organise and manage the workforce – a workforce constituted as such, to ensure its pliability and productivity. As Dickens illustrates, the philosophy of 'Fact' and factory denies Coketown's workers the opportunity to flourish in her or his own unique way. Quite literally, it denies them their individuality, and the workers' own emerging response – a response based on a collective identity – is, Dickens contends, equally coercive and no less mistaken.

What does it therefore mean to flourish? Here, a new materialist emphasis on the particularity of 'intra-action' (Karen Barad's neologism for '*the mutual constitution of entangled agencies*'; italics in the original; 2007, 33) – and on the micro-politics of lived experience – overlaps with a capabilities approach to environmental justice (Holifield, Chakraborty, and Walker 2018, 4). Developed by Amartya Sen and Martha Nussbaum (see Nussbaum 2003), a capabilities approach is focused less on resources (the distribution of environmental harms or goods, for example) than on an individual's opportunities to find meaningful fulfilment.[4] Whilst this does not rule out the macro-political dimension of structural forces – the material-discursive convergence of industry, capital, and modernity whose impact can be traced in Coketown itself – it does offer a useful analytical tool for engaging with Dickens's own response to the impact of industrial modernity: the concept of

'Fancy', a term he used 'interchangeably with "imagination"' (Slater 2011, 14). In Dickens's work, 'Fancy' is a powerful marker of a free and self-creating life, itself the over-arching goal of a capabilities approach (Nussbaum 2006, 74); by extension, it also forms the basis of a flourishing society. '[A] nation without fancy', Dickens wrote in *Household Words*, 'never did, never can, never will, hold a great place under the sun' (1853, 97; see also Slater 2011, 18). This argument is written into the discursive fabric of the novel itself, which departs from the Victorian realist norm in its defamiliarising, 'fanciful' approach to its subject matter; as I discuss in this chapter, that approach encourages the reader to participate in a process of imaginative and ethical extensionism that breaks down not only class but also gender and species divides.

This imaginative leap encompasses the story of Gradgrind's daughter, Louisa, victim of a kind of gendered environmental injustice, and takes substantive, material form in the shape of the circus. As Michael Schlicke points out, the circus is 'the central repository of human value in *Hard Times*' (1985 [1988], 143), and its presence both bookends the novel and defines Coketown's antithesis. As 'a creative alternative space of becoming', to use Rosi Braidotti's phrase (2011, 7), the circus embodies what posthuman scholars have defined both as a nomadic sensibility (Baker 2018) and a nomadic subjectivity (Braidotti 2011). It is at once the product of the need for entertainment – for the flights of 'Fancy' it offers its audiences in Coketown's otherwise unrelenting world of work – and, more subtly and more profoundly, of the self that exceeds self, a 'relational self that functions in a nature–culture continuum' (Braidotti 2018, 341). Embattled as the circus may be, in the new world of industrial modernity's making, it represents a kind of ark, fluid, flexible, and fair-minded, a utopian project (processual and performative) that embodies what Braidotti has described as 'the desire, the ability, and the courage to sustain multiple belongings' (2011, 10). Consequently, it is the circus girl, Sissy, who emerges as the novel's heroine, 'a moral beacon throughout the book' (Schlicke 1985 [1988], 143), in whose relational view of human needs Dickens locates a powerful alternative to the demands of a self-interested society.

*

In *Hard Times*, Dickens personifies the complex matrix of forces that were then transforming society – or standing by as it was transformed. As Dickens recognised, this transformation would not have been possible without the involvement (or complicity) of the old, established order, here represented by the louche figure of James Harthouse, quintessence of the irresponsible dilettantism indicted by Thomas Carlyle and discussed in Chapter 1.[5] Harthouse stands in distinct contrast to those energetic apostles of industrial modernity, Josiah Bounderby, 'banker, merchant, manufacturer' (Bk. I, Ch. 4, 19), and his ally, Thomas Gradgrind, now 'virtually retired from the wholesale hardware trade' (Bk. I, Ch. 3, 14), and intent on advancing his ideas by seeking a place in Parliament. Their alliance underpins society's transformation, a marriage of convenience made flesh in what amounts to Gradgrind's gift (or sacrifice) of his daughter Louisa as wife to his inexecrable colleague, that 'Bully of humility' (Bk. I, Ch. 4, 19), Bounderby. As his name suggests, Bounderby is 'a vain blusterer' (Bk. III, Ch. 9, 270) who simply cannot be trusted, least of all when he claims to be made a self-made man: in Dickens's fiction, childhood stands for a kind of unimpeachable purity, but Bounderby tarnishes the memory of his own by putting in its place an invented tale of abandonment and suffering overcome by sheer force of will. Much rests on that tale: it justifies Bounderby's insistence that individualism is sufficient – that the isolated individual can thrive in a dynamic and confrontational new society.

In Bounderby, Dickens combines the interlinked forces of industrialisation and capitalism, each dependent on the other. In the world shaped by this combination, place and people are alike exploited, 'as something wholesale, of which vast fortunes were made' (Bk. II, Ch. 6, 149). But as Dickens just as pointedly insists, there is at work a powerful and compelling philosophy – an 'infinite grinding at the mill of knowledge' (Bk. I, Ch. 9, 57) – that legitimates the crude influence of industrial capitalism, and that philosophy is embodied in Gradgrind. His theories have their roots in the Enlightenment emphasis on the rational reformulation of human behaviour, but the immediate source of Gradgrind's thinking is what had then become a kind of philosophy-in-office, quietly ruling society's horizon of expectations. Dickens variously refers to 'Utilitarian economists' and 'Commissioners of Fact' (Bk. II, Ch. 6, 153), but his target is, more exactly, a concatenation not just of Utilitarianism, the consequentialist philosophy set out by Jeremy Bentham, but of political economy, a form of liberal economics derived from Adam Smith's *The Wealth of Nations* (1776), and a singularly pessimistic view of human nature heavily influenced by Thomas Malthus's views on population growth; Malthus had argued that only starvation would keep a rising population in check, an argument from which David Ricardo derived his 'Iron Law of Wages' (Wilmer 1985, 19–21). The resulting fusion was a mechanistic, materialistic philosophy which, as elaborated by James and later John Stuart Mill (Wilmer 1985, 20–1), placed the emphasis on economic motives as drivers of human behaviour, which it sought to understand in rational but wholly self-interested terms. Equipped with this understanding of human nature, contemporaries set to work building a better society that was itself rational, uniform, and efficient, ruled by matters of fact. 'Facts alone are wanted in life', Gradgrind declares (Bk. I, Ch. 1, 7), but figures are equally important, especially when a pound sign is involved. This world is dominated by what Carlyle called the cash 'nexus' (1843 [1897], Vol 6, 379), which reduces all social (or socio-natural) relationships to a question of economic gain; 'and what you couldn't state in figures', declares Dickens's narrator, 'was not, and never should be, world without end' (Bk. I, Ch. 5, 27).

> It was a fundamental principle of the Gradgrind philosophy that everything was to be paid for. Nobody was ever on any account to give anybody anything, or render anybody help without purchase. Gratitude was to be abolished, and the virtues springing from it were not to be. Every inch of the existence of mankind, from birth to death, was to be a bargain across a counter. And if we didn't get to Heaven that way, it was not a politico-economical place, and we had no business there.
>
> (Bk. III, Ch. 8, 264)

Gradgrind, then, is not just a social theorist and aspiring politician: he is the embodiment of a philosophy that is doubly dangerous because it conceives of itself as high-minded and moral. There is, however, a further element to Gradgrind's thinking, which derives directly from the mechanistic, individualistic, and reductionistic nature of Enlightenment thinking. As Adorno and Horkheimer insisted, Enlightenment thinking first separates out 'man' from the world of mere matter, and then seeks to instal him as master over the material world he once feared (1997 [2010], 3–6). This has convenient consequences for industrial capitalism, which seeks cheap resources to exploit: a disenchanted natural world, once separated out as such – as 'Nature' – can then be discounted as an economic externality (Altvater 2016, 149). Although that world is intrinsically valuable, because integral to what Jason W. Moore calls the 'web of life' (Moore 2016, 7), no (monetary) value is attached to it. Since the cash nexus determines that this is the only way in which value *is* measured, 'Cheap Natures' (Moore 2016, 2) (cheapened and debased) are the result.

In Western thinking, adds Val Plumwood, this process of hyperseparation (1993 [1997], 49) – of marking out nature and the natural as separate and inferior – is linked to a range of other dualistic devices that establish dominant paradigms of superiority and inferiority; these include gender and class, both of which play a prominent part in Dickens's narrative. This process of diminution – of 'othering' – also finds its correlative in the second notable consequence of the philosophical position that Gradgrind espouses: his assumption that society is simply the aggregate of atomised individuals, who can themselves be understood in simplistic, mechanistic terms, as facts in a factful world. It is, as Amartya Sen observes, a 'miniaturisation of human beings' (2006 [2007], xiii); it represents a willed ignorance of all those many things that constitute human happiness but are somehow not reducible to facts and figures.

A culture that celebrates individualism is, moreover, one that ignores the structural forces with which it is complicit, such as the operations of capital, or the way in which industrial modernity constructs the isolated individual as an interchangeable unit in a machine-like construct, a 'universal machine … with its related machine-like qualities of abstract value, sucking away at the humans on the levels beneath' (Morton 2013, 5). But by reducing Coketown's workers to the status of interchangeable units, subjects of Gradgrind's own coldly logical (mis)understanding of their nature and needs, Gradgrind has helped to create a body of people who behave as such: all (mis)treated in the same way, their 'unique and choiceless' but shared identity (Sen 2006 [2007], xv) becomes a source of strength, of solidarity, opening up the possibility of a revolutionary escape from their collective predicament. Thus, the paradoxical product of Gradgrind's philosophical faith in individualism is the inconvenient emergence of its antithesis: class, and class conflict.

Like many middle-class writers of the time, Dickens baulks at the prospect of violence; nonetheless, he identifies the reasons why it increasingly appeared inevitable. Violence begets violence, but violence takes different forms. As Sen argues, the very act of imposing a single, monolithic, 'solitarist' (2006 [2007], xii) definition (such as class) on '*diversely different*' identities (italics in the original; xiv) makes mutual incomprehension and conflict more likely, and might itself be regarded as an insidious but palpable kind of violence. The point, as Slavoj Žižek has argued, is that violence takes two main forms: subjective violence, or 'violence performed by a clearly identifiable agent' (or subject) (Žižek 2008 [2009], 1), and its less visible twin, objective violence. Objective violence, Žižek argues, manifests itself in two ways. First, there is the symbolic violence contained within language itself, 'the relations of social domination reproduced in our habitual speech forms' (Žižek 2008 [2009], 1), of which the designation of Coketown's mill-workers as 'Hands' (Bk. I, Ch. 10, 64) is a compelling example. In the broader sense of symbolic violence as the 'imposition of a certain universe of meaning' (Žižek 2008 [2009], 1), it can also be detected in the habitual way in which middle-class writers such as Elizabeth Gaskell (see, for example, Gaskell's *Mary Barton*; 2006 [2008], 184–5) and Dickens (Bk. II, Ch. 4, 131–6) presented collectivist bodies such as trade unions as either ignorant and misguided, or worse, as violent, sinister and secretive (Hall 2017, 63). The second form of objective violence, argues Žižek, is systemic: it encompasses 'the often catastrophic consequences of the smooth functioning of our economic and political systems' (2008 [2009], 1). But because these forms of objective violence have become normalised – naturalised within the apparently 'peaceful state of things' – it is subjective violence which 'is seen as a perturbation of the "normal"' (Žižek 2008 [2009], 2). Moreover, objective violence extends to the existence of injustice itself: as Rob Nixon argues, 'a widening gulf between rich and poor is a form of covert violence in its own right that is often a catalyst for more recognisably overt [i.e. subjective] violence' (2011, 10–11).

Here, we can introduce a further concept, which is itself useful in defining the nature of the objective, systemic violence to which Žižek refers: slow violence. As Nixon defines it, slow violence is 'incremental and accretive' (2011, 2), 'pervasive but elusive' (3), a violence 'dispersed across time and space' (2). It is a violence that is difficult to identify, imagine, think, or represent: '[s]tories of toxic buildup, massing greenhouse gases, and accelerated species loss due to ravaged habitats are all cataclysmic, but they are scientifically convoluted cataclysms in which casualties are postponed, often for generations' (Nixon 2011, 3). Nevertheless, there *are* casualties, and the human casualties are, typically, the poor (Nixon 2011, 4). 'It is against such conjoined ecological and human disposability' that an 'environmentalism of the poor' is directed, Nixon adds (2011, 4), and against which a broadening environmental justice movement is itself struggling (4–5).

How do these forms of violence – these injuries and injustices – manifest themselves in the material reality of Coketown and its environs? How, in turn, does the novel construct its response to that reality? In place of a living, adaptive, and sustainable environment, industrial modernity has shaped a dead and deadening one. Coketown is at its centre, a necropolis 'of machinery and tall chimneys', with 'a black canal', 'and vast piles of building ... where there was a rattling and a trembling all day long' (Bk. I, Ch. 5, 26). This bleak whole is hidden under a mass of darkness 'impervious to the sun's rays' (Bk. II, Ch. 1, 105): even on a fine, midsummer day, notes the narrator, 'Coketown lay shrouded in a haze of its own' (Bk. II, Ch. 1, 105). In this gloomy citadel, environmental injustice is at its most pronounced

> at the heart of the labyrinth of narrow courts upon courts, and close streets upon streets ... where the chimneys, for want of air to make a draught, were built in an immense variety of stunted and crooked shapes, as though every house put out a sign of the kind of people who might be expected to be born in it.
>
> (Bk. I, Ch. 10, 64)

Dickens is making a serious point: '[t]he [Victorian] urban poor did tend to be small, slightly built and pallid of skin', notes Carl Chinn (1995 [2006], 89). 'They were made that way by their inadequate diet, by the polluted environment in which they had to live and by their dreadful working conditions' (Chinn 1995 [2006], 89). This was the impact of slow violence, enacted over a lifetime, and written into the body; a new, social body, to paraphrase Mary Poovey's account of the way in which the labouring poor were increasingly represented as an 'undifferentiated whole' (Poovey 1995, 4). Lungs grew 'more bronchitic and cheeks paler' (Clapp 1994, 38); 'rickets caused by a lack of sunlight' itself became an urban phenomenon (Daunton 2000, 62).

Yet this is, as Dickens's narrative insists, 'the hardest working part of Coketown' (Bk. I, Ch. 10, 64). Environmental injustice has become environmental classism, as Karen Bell has defined it (2020, 3). The inhabitants of these courts are neither unemployed nor unemployable. Far from being part of the 'residuum' that middle-class writers wrote off as a kind of unfortunate but inevitable by-product of modernity, these 'hard-worked' people (Bk. I, Ch. 10, 64) are able-bodied – able-bodied, that is, whilst they remain watchful and alert amidst the 'crashing, smashing, tearing pieces of mechanism' that constitute an often lethal workplace (Bk. I, Ch. 11, 69) – and in full-time work, which is to say that they have little time for anything but work, a point that Dickens also raises repeatedly. 'You saw nothing in

Coketown', the narrator observes, 'but what was severely workful' (Bk. I, Ch. 5, 25). Coketown is, in other words, the expression and logical culmination of the Victorian 'cult of work': 'men worked, women worked, children worked, and if they could not work, they went to the workhouse' (Matthew 2000, 5). 'By the middle of the nineteenth century', notes Hans-Joachim Voth, 'working hours had reached levels that were probably higher than ever before or since' (2004, 270).

In this 'wilderness of smoke and brick' (Bk. I, Ch. 14, 88), hunger is not the issue, although economic downturns might make it so. Bounderby is, as usual, guilty of self-serving exaggeration when he claims that his 'Hands' have but 'one ultimate object in life … to be fed on turtle soup and venison with a gold spoon' (Bk. II, Ch. 2, 120). The problem is rather the smoke they consume, and according to Bounderby, relish eating. '[O]ur smoke', he tells Harthouse, is 'meat and drink to us': '[i]t's the healthiest thing in the world in all respects, and particularly for the lungs' (Bk. II, Ch. 2, 119). Thus, the slow violence of Coketown's 'killing airs and gases' (Bk. I, Ch. 10, 64), inevitable concomitant of its fossil-fuel-powered productivity, is neutralised through the language of obfuscation, even of outright denial; it is absorbed by 'habitual speech forms' (Žižek 2008 [2009], 1) that serve to conceal subjective violence. These forms of obfuscation and denial are all the more unconscionable because already (as arguments) untenable. Dickens was understandably vague about the exact nature of those 'airs and gases' because, as Bill Luckin points out, it was not until the 1870s that research begun to focus on the human health impacts of polluted air (2015, 15), but the early Victorians were already aware that 'smoky factory chimneys and domestic hearths undermined health – particularly among the very young and the elderly – for miles around' (14). As Lord Palmerston was then arguing, the problem was rather the stonewalling, obfuscation, and outright denial of 'smoke-producing monopolists' (*Hansard* HC Deb., 8 August 1853), of whom Bounderby is a representative example.

Pollution is not, however, the only form of violence that this factory town enacts: subtle denials also constitute a kind of violence, and in Coketown, those denials are encoded in its very fabric. The constitutive loss of contact with the nonhuman and natural world (both biotic and abiotic) – a theme to which I return later in the chapter – is one such denial, but there is another, which relates to a further form of erasure in the built environment. In *The Seven Lamps of Architecture* (1849), the critic John Ruskin argued that buildings tell stories of human intra-action and involvement, developing over time, stories that signify the value and importance of those buildings as part of communities of care that extend over the generations (Cook and Wedderburn 1903–12, Vol. VIII, 233–4). But towns such as Preston, on which Dickens modelled Coketown, expanded in size at such a rate that whole districts seemed to spring up overnight: 'in the innermost fortifications of that ugly citadel', as the narrator says of Coketown, the streets and courts appear to have 'come into existence piecemeal, every piece in a violent hurry for some one man's purpose' (Bk. I, Ch. 10, 94). In so far as these *ad hoc*, jerry-built structures have any purpose, it is to make a profit, in accordance with the 'fundamental principle of the Gradgrind philosophy' (Bk. III, Ch. 8, 264). Where the buildings have been more carefully contrived, they have created a built environment of unrelenting monotony and 'direful uniformity' (Bk. I, Ch. 14, 88); they are concrete proof of the 'severely workful' philosophy of 'Fact' to which the town is sacred (Bk. I, Ch. 5, 27), whether it is the 'lying-in' hospital, Gradgrind's model school, or one of many chapels, each 'a pious warehouse of red brick' (Bk. I, Ch. 5, 26). All are alike, the narrator points out: 'the jail might have been the infirmary, the infirmary might have been the jail, the town-hall might have been either' (Bk. I, Ch. 5, 27); and each is an imposition, alien and alienating.

The townscape is, Dickens implies, unrelated to those who live there, at once a potent illustration of society's disregard for its working majority, and a mechanism for their segregation and social control: in its very fabric, the town enacts (or acts out) forms of violence. Yet this is not felt as such. The systematic exploitation of a wage labour force requires more than its dependency, because dependency is in itself insufficient: when the workforce acts as one – when it works to support itself – collective action is still possible. Exploitation requires compliance: it requires a workforce that shares the same values as its masters, and internalises the problems generated by change. New chapels, churches, and schools (such as Gradgrind's model school) all form part of that attempt to impose this 'universe of meaning' (Žižek 2008 [2009], 1). Similarly, the effective exploitation of the workforce requires its standardisation, which also requires its alignment with standardised days and shifts. Thus, '[t]ime went on in Coketown like its own machinery' (Bk. I, Ch. 14, 88), marshalled by 'a deadly-statistical clock ... which measured every second with a beat like a rap upon a coffin-lid' (Bk. I, Ch. 15, 93).

In each of these respects, therefore, industrial modernity has shaped a hostile and a coercive environment, alienating by accident and design, which is still more inescapable because it is presented as a fact through the language *of* 'Fact'. The language of 'Fact' declares itself to be reasoned and rational, thereby justifying the shape and nature of this new but profoundly divided society, but its method is itself reductionistic not holistic, and its studied air of neutrality is simply a mechanism for normalising what is abnormal and abnatural (Taylor 2016, 5) and above all unjust. After Žižek, it too forms part of the insidious landscape of objective violence that distorts the lives of Coketown's 'Hands' (Bk. I, Ch. 11, 69) – a term that itself captures the workers' interchangeability, their loss of individuality, and in a very real sense, their humanity. What matters is their usefulness – their utility – and that usefulness is captured in the reductive language that employers use in relation to them.

For Dickens, therefore, the language of 'Fact' leads only to the 'parliamentary cinder-heap', the 'national dust-yard' (Bk. II, Ch. 9, 184); it culminates, all too literally, with the dust-heaps he would describe in his final novel,

Our Mutual Friend, a landscape made up not just of cinders and ash but of those discarded and disvalued people who are reduced to picking it over in search of something (anything) that might make a penny or two. By contrast, Dickens insists, language is (or should be) an 'unruly superfluity' (Beer 1985, 48), like life itself – and in *Hard Times*, that superfluity transforms factories into 'Fairy palaces', machines into 'melancholy-mad elephants' (Bk. I, Ch. 11, 69), and, perhaps most memorably of all, trails of smoke into 'Smoke-serpents' (Bk. I, Ch. 11, 69). The imaginative excess of Dickens's description is very much to the point; it forms part of his critique, and part of his constructive counter-proposal. For Dickens, it is not enough to rely on realism to represent his subject matter, since realism is, in its pursuit of facticity, complicit with the agencies that are quietly conspiring against a flourishing existence: like the language of 'Fact', it presents itself as a mirror held up to the world, as a kind of unimpeachable witness to what it represents. The language of 'Fact' must be met by the language of '*Fancy*', as itself a profoundly important imaginative tool for destabilising the apparent certainties on which this new world is raised, and defamiliarising the forms of violence that 'Fact' has normalised.

Dickens's reference to Coketown's 'interminable serpents of smoke' (Bk. I, Ch. 5, 26) – its 'monstrous serpents of smoke' (Bk. I, Ch. 11, 69) – is a case in point. Bounderby might well deny that smoke is a risk to human health, but it constitutes a form of slow violence: to describe these trails of smoke as interminable serpents is, therefore, to capture both the hypnotic, nightmarish quality of smoke-without-end, ever spiralling from Coketown's sea of chimneys, and its insidious, threatening quality; it conjures up the image of Eden's sinister intruder, the serpent in the garden (Genesis 3:14), to which the narrative elsewhere and explicitly relates them (Bk. I, Ch. 11, 70). When Dickens invokes those 'Smoke-serpents' (Bk. I, Ch. 11, 69) (as he does six times in the novel), however, his intention is not simply to challenge the faith of all those who saw in smoke a marker of the nation's new prosperity, and to substitute for it the makings of a 'toxic sublime' (Taylor 2016, 47). He means something more: he means us to understand that the material world *is* alive and lively, and only 'Fancy' (the imagination) can supply the deficit between what the philosophy of 'Fact' – and the Enlightenment's epistemological project – would have us believe it is, and an agential reality. Imagination enables Dickens to grasp what new materialist theory makes plain: the liveliness and dynamism (and hence unpredictability) of the material world, in all its manifestations.

Here as elsewhere, Dickens's presentation of that agentiality reflects his argument with rationalistic and reductionistic interpretations of reality as mere matter, as mechanical (Levine 1981 [1983], 19). There is a further dimension to the narrative's insistence on these serpents of smoke. They are products, notes Dickens, of a forest of chimneys, and like the mines that dot the countryside around Coketown (Miller 2021, 4–7), those chimneys mark the decisive transition from what Miller calls 'an organic [to] the world's first extraction-based economy' (2021, 6). What may stick in the reader's mind, however, is Dickens's comparison of those chimneys to 'competing Towers of Babel' (Bk. I, Ch. 12, 79), a reference to the Biblical story of Noah's descendants, who sought to reach heaven, and who were punished for their presumption by having their languages reduced to a babble, to confused nonsense. For Dickens, that babble is the language of 'Fact', or of what the narrative elsewhere describes as 'blue book, report, and tabular statements A to Z' (Bk. I, Ch. 9, 57–8), discursive expressions of the 'reasoning powers' (Bk. I, Ch. 14, 89) whose reductive thinking he indicts. For Dickens, this exclusive reliance on the measurable and the calculable will lead only to a still greater reliance on their concomitant: systems, processes, organisations, and institutions, all both uncomprehending and incomprehensible. This way lies the hyperobjects that Timothy Morton (2013) describes as industrial modernity's corollary, vast entities that (like the Court of Chancery in *Bleak House*) may be our creations but cannot now be controlled or even fully comprehended. Facts are, for Dickens, part of a system of thought that perpetuates systems, and Dickens is never more Carlylean in doubting whether, in an 'Age of Machinery, in every outward and inward sense of that word' (Carlyle 1829 [1897], Vol 14, 465), that is or could ever be a good thing (Slater 2011, 11).

As this discussion highlights, what *Hard Times* offers the reader is a 'fanciful' narrative method, which (as Dickens put it) eschews the 'literal and catalogue-like' (qtd in Forster n.d., 563), and substitutes for it a mode that simultaneously defamiliarises everything that is outwardly familiar, investing it with a strange and sometimes grotesque life of its own (Slater 2011, 15–16). This mode is, in other words, the literary embodiment of 'Fancy' itself, as a capacity for wonder and for imagination. In *Hard Times*, however, 'Fancy' takes on a still deeper meaning, since its denial by 'Fact' is an important theme within the novel; as a capabilities approach infers, and Nussbaum's work underlines, the freedom to wonder, imagine, and (self-) create is itself integral to a full and flourishing life. It is important not least because, as Dickens's narrative emphasises, 'Fancy' refers to the imaginative ability to empathise with the other, and conceive of otherness, in terms that both span the species divides of human and nonhuman *and* encompass an agential reality; indeed, '[b]eing able to live with concern for and in relation to animals, plants, and the world of nature' is one of the core capabilities that Nussbaum identifies (capability 8; 2006, 77). 'Fancy' enables us to take joy in this world, but also, and as Dickens does through his depiction of the circus, to imagine its alternate: as we shall see, it is the circus – at once marginal to the world the novel describes, yet central to the narrative – that enables Dickens to conjure up modernity's antithesis, as a space within which fellow-feeling extends beyond self and species, and life can (in spite of its many vicissitudes, and in often unconventional ways) flourish.

'Fancy' is, in other words, central to Dickens's conscious and deliberative working out of an anarchic, spontaneous response to the impositions of industrial modernity, with its inequalities and inequities, and central to his own estimation of the value of imaginative literature '[i]n an utilitarian age' (Dickens 1853, 97). As Dickens observes, '[f]orbearance, courtesy, consideration for the poor and aged, kind treatment of animals; abhorrence of tyranny and brute force – many such things have been first nourished ... by this powerful aid' (Dickens 1853, 97). '[F]airy-tales, fictions, and fables' (Dickens 2003, Ch. 21, 333) all have the power to cultivate these capacities for understanding, these essential capabilities, a point that Nussbaum herself takes up in her reading of *Hard Times* in *Poetic Justice: The Literary Imagination and Public Life* (1995). The very act of immersing oneself in a book, she contends, involves an imaginative and ethical extension of the self, paralleled in the novel by Dickens's emphasis on precisely those qualities (such as the imagination) that an Enlightenment rationality leaves out of its formulation of a just, fair, and flourishing society (Nussbaum 1995, 1–2).[6]

Problematic as Nussbaum's approach may be (see, for example, Posner's critique of it; 1997), her point echoes Dickens's own: without the opportunity to cultivate the imagination, and with it, this crucial sense of empathy, lives are necessarily stunted and deformed. In *Hard Times*, Dickens illustrates his point through the story of Louisa, whom we encounter as she and her brother sneak a look – to Gradgrind's consternation – at the circus in performance. (Gradgrind immediately suspects that an 'idle story-book' may be at the bottom of this otherwise inexplicable behaviour; Bk. I, Ch. 4, 23). Although only a teenager, Louisa appears exhausted, beaten down, yet 'struggling through the dissatisfaction of her face, there was a light with nothing to rest upon, a fire with nothing to burn, a starved imagination keeping life in itself somehow, which brightened its expression' (Bk. I, Ch. 3, 17). Her capacity to wonder has been stifled, almost (but not quite) extinguished, exhausted as she now feels herself to be:

> 'I have been tired a long time', said Louisa.
> 'Tired? Of what?' asked the astonished father.
> 'I don't know of what – of everything, I think'.
>
> (Bk. I, Ch. 3, 17–18)

As the story unfolds, Dickens brings to bear his fanciful narrative mode to deepen our sense of Louisa's predicament, and the risk it poses – to herself, to those near her, to a society that depends on the strictly regulated control of desire and especially of female agency: as Louisa tells her father, when he communicates to her Bounderby's proposal of marriage, the chimneys of Coketown seem to pour forth nothing 'but languid and monotonous smoke. Yet when the night comes, Fire bursts out, father!' (Bk. I, Ch. 15, 96). Louisa's fire-gazing signifies the unanswered needs that still exist within her, simply waiting (as the predatory Harthouse recognises) for the chance to erupt, threatening to bring shame down on her; yet the real shame is that she has been more or less fashioned to marry a man like Bounderby by a father blinded to the 'subtle essences of humanity' by his own 'unbending, utilitarian, matter-of-fact' outlook (Bk. I, Ch. 15, 96). It is because he does not pretend 'to anything fanciful, fantastic, or (I am using synonymous terms) sentimental' that he cannot see how short-sighted that outlook is, or recognise that Bounderby does indeed do her an injustice (Bk. I, Ch. 15, 95): that the women Gradgrind believes her to be is, as she herself insists, little more than a child in every sense but the physical – if ever she had a childhood. 'What are my heart's experiences?' she demands of Gradgrind (Bk. I, Ch. 15, 97).

> 'What do I know, father', said Louisa in her quiet manner, 'of tastes and fancies; of aspirations and affections; of all that part of my nature in which such light things might have been nourished? What escape have I had from problems that could be demonstrated, and realities that could be grasped?' As she said it, she unconsciously closed her hand, as if upon a solid object, and slowly opened it as though she were releasing dust or ash.
>
> (Bk. I, Ch. 15, 97)

Dust or ash: one is reminded, once again, of the by-products of the extractive ecology (to paraphrase Miller 2021) that the new Coketowns of the country were generating and by which they were constructed, an ecology that itself does violence to its subjects. In the case of well-to-do women such as Louisa, however, that violence – that injustice – has been naturalised within a language that presents itself as logical, even deferential, and so it is Louisa's behaviour which comes to seem errant, scandalous, and self-destructive, as itself 'a perturbation of the "normal"' (Žižek 2008 [2009], 2). Indeed, the injustice of her treatment may itself be read as a form of environmental injustice. As environmental justice scholar Rachel Stein explains,

> [w]hen ... we view our bodies as 'homes', 'lands', or 'environments' that have been placed at risk, stolen from us, and even killed due to social or physical harms that may be exacerbated due to our gender and sexuality – we may understand the need for new perspectives on environmental justice.
>
> (2004, 2)

Louisa's fate (and her reaction to it) is anticipated in the scene where, whilst she is still young, Bounderby insists on giving her a parting kiss on the cheek (Bk. I, Ch. 4, 24–5). Louisa senses what she might not even be able to describe – that the kiss he plants on her has taken on a rapacious, sexual dimension – and she tries furiously to rub away every trace of it (Bk. I, Ch. 4, 25). In turn, we can see how Louisa's marriage is itself an invasive appropriation of her body, and a substantive denial of her rights over that body; in accordance with the domestic ideology of the mid-Victorians (Barker and

Chalus 2005, 5), she will now be expected not only to share Bounderby's bed, but to bear his children, and bear the risk of having them. If her marriage to him seems to be the logical, the rational choice, then a desire to become a mother is assumed to follow, naturally; but on the one hand, Louisa's nature has been grotesquely deformed by her upbringing, leaving her with no real sense of what an embodied existence might be; and on the other, 'nature' has been constructed as culture's inferior obverse, with which she has herself been aligned by a patriarchal society. Louisa is, in other words, trapped by these interlocking systems of 'male privilege' (Stein 2004, 9), which position her as useful because of her role as wife and potential mother, but inferior by virtue of a gender that positions her as closer to nature (Harvey 2005, 81).

The effect of that entrapment on Louisa is brought home in a later confrontation with Gradgrind, when she makes it painfully clear to him that the life he has given her (ashes to ashes) is in fact a denial of life:

> How could you give me life, and take from me all the inappreciable things that raise it from the state of conscious death? Where are the graces of my soul? Where are the sentiments of my heart? What have you done, O father, what have you done, with the garden that should have bloomed once, in this great wilderness here!
>
> (Bk. II, Ch. 12, 200–1)

Louisa puts her hand to her heart: the wilderness within her parallels the abnatural wilderness of brick and smoke without; they are both the sterile products of the same reductive, dualistic mindset, the same supposed philosophy of 'Fact'.

'Artists act as Cassandras', argues Stein: they help us understand 'the complex and often conflicted positions' that women occupy, and

> inspire us to hope for a truly just world by helping us envision positive resolutions to current disasters and the possibility of forming creative, non-traditional alliances that recreate an enlarged sense of 'home', 'family', or even 'body' beyond current divisions and borders.
>
> (Stein 2004, 13)

Dickens's own sense of how those alliances might be constructed in a new and more expansive fashion is reflected in his positioning of Sleary's circus as an alternate framework to conventional society: as I discuss in more detail, below, Sleary's circus community is decidedly non-traditional, in ways that subvert the domestic ideology of the day, and transcend the species divide that increasingly separated human and nonhuman animals.

Once again, it is through the fanciful that Dickens introduces his subject matter. Early in the novel, Bounderby and Gradgrind encounter the circus folk, who are quartered, significantly enough, in a hostelry called the Pegasus's Arms, a reference to the winged horse of Greek legend (Bk. I, Ch.

6, 32). Here, they meet Mr E. W. B. Childers, who is short of leg, broad of chest, and wide of gait, rather as if he has grown up on horseback, as many of the circus performers seem to have done (Bk. I, Ch. 6, 37; see also Bk. III, Ch. 7, 258–9). Indeed, 'The Flying Childers' was the name of a then-famous circus horse (Dickens 2006 [2008], 288, footnote 33). With this kind of pedigree, it is no surprise that Mr Childers smells variously 'of lamp-oil, straw, orange-peel, horses's provender, and sawdust; and looked the most remarkable sort of Centaur, compounded of the stable and the play-house. Where one began, and the other ended, nobody could have told with precision' (Bk. I, Ch. 6, 33).

It is a fanciful description, quite at odds with the language of 'Fact', which has already supplied the reader with a description of a horse that reduces it to its component parts (Bk. I, Ch. 2, 9–10), and leaves the reader to supply it with life (the narrator returns the compliment by describing Gradgrind in a similarly atomistic way, as a collection of body parts in loose constellation [Bk. I, Ch. 1, 7]). Manifestly, this description breaks with the rules of all that is reasonable, since Childers is manifestly not a horse, nor a part of one, no matter how much he might smell of the stables. It also assumes that the reader is susceptible to this kind of suggestive writing, which simultaneously transgresses the inviolable boundary of human being, whilst also invoking a rich literary tradition of mythological creatures that casually disregards what is probable or possible (winged horses, horses-as-men). But then Childers is a material-discursive formation, whose line of work is an enactment of this performative identity; every day, he becomes 'the Wild Huntsman of the North American Prairies' (Bk. I, Ch. 6, 33), a narrative which requires physical daring, a (very) well-trained horse, and a great deal of imagination on the part of the audience, which is also required to believe that a grown man is Childer's infant son.

To Gragrind's surprise, this is precisely what the circus audience does supply, willingly, as does the reader, as she or he reads Dickens's description of Childers as a curious kind of hybrid, somewhere between a horse and a man, and at once a part of nature, and of culture. The combination is a critical one. We are today accustomed to thinking of the circus as a questionable proposition, associated with cruelty to captive animals; indeed, Nussbaum cites a historic case in recent Indian law, whereby circus animals were given legal protection to ensure their own dignity and well-being (Nussbaum 2006, 321–2). In this reading of it, the circus has simply become a part of the historical (and historic) shift to which John Berger referred in *Why Look at Animals* (1977) (Berger 2009, 12–13), as working relationships with animals were supplanted by new forms of mechanisation, such as the horseless carriage (22), and animals instead become objects at which to stare, in zoos – or circuses – everywhere 'rendered absolutely marginal' (35–6). But this is not the world that Dickens describes. In one key respect, he barely describes it at all. We never see the circus in performance: the closest we come to it is the sight of Tom and Louisa (Bk. I, Ch. 3, 16–17), noses pressed to a fence, trying to snatch a glance at the circus entertainments denied to *them*. What we are afforded, instead, is an insight into a working world, the world of 'Sleary's Horse-Riding' (Bk. III, Ch. 7, 257), which is centred on the relationship between horse and rider (each increasingly inseparable from the other). Audiences might spectate, as Berger suggests we all now do as we look at animals, but this is not their world: the men and women who make up Sleary's circus work alongside their animals, and share the risks with them, creating a kind of (ethical) equivalence that makes fascinating sense of Sleary's own pronouncement:

> I don't pretend to be of the angel breed myself, [but] good tempered or bad tempered, I never did a horthe a injury yet, no more than thwearing at him went, and that I don't expect I thall begin otherwithe at my time of life, with a rider.
>
> (Bk. I, Ch. 6, 41)

Sleary's point of reference, in other words, is the nonhuman animal, a moral compass that is then extended to the human – not the other way around. Inverting humanist arguments for ethical extensionism, whereby animals are given moral consideration to the extent that they share human characteristics or attributes (such as sentience), Sleary simply equates riders with animals. This is not to argue – any more than he does himself – that Sleary is an infinitely indulgent and benign ring-master: 'I don't thay but what, when you mith'd your tip [i.e. fell during a performance; Dickens 2006 [2008], 289, footnote 34], you'd find me cut up rough, and thwear an oath or two at you' (Bk. I, Ch. 6, 41); the point is simply that, in a nondualistic working world defined by the relationalities (rather than the entities) that constitute it, there is neither time nor inclination to impose a humanist hierarchy or an exceptionalist logic that separates out the humans as superior.

As the reader engages with this sense of cross-species co-dependency, she or he may wonder whether Sleary's perspective has rather more general applicability – the point that Nussbaum makes, in *Poetic Justice* – and wonder at the limits of her or his own ethical horizon. Today's society is still dependent on animals, yet that dependency itself entails unprecedented forms of intervention, exploitation, and subjection (Derrida 2008, 25), in apparently casual disregard of the denials (of well-being, of dignity) that it assumes. But this cross-species intra-action (at once processual and performative) is not the limit of the world that Sleary's circus encircles. Imagination enables us to think outside ourselves, Dickens suggests; to recognise that the self is never in any case synonymous with the self, but always already entangled with others, with otherness. This has equally important consequences for the way in which the human members of Sleary's community interact, confounding the logic of Gradgrind's class-divided, self-interested world, and the gendered, domestic ideology towards which Dickens gestures in describing Louisa's fate.

A relational ontology – a relational way of being, mindful of connectivity and co-constitution – is also a flat ontology, in which hierarchies are trumped by relationships, and a fixed order by a processual fluidity. It follows that, in Sleary's circus, his rule is a nominal one; what matters is a functioning whole, towards which everyone contributes equally – a logic which also has its casualties, as Sissy discovers when her father, Signor Jupe, decides that it is better for her and for the circus if he leaves it. Contemporaries would have seen in the circus an extended family of the kind that many, as recent immigrants to the city, remembered from their rural upbringing, where poverty was in part offset by 'a tradition of small community support' (Howe 2009, 9); it was this that city-life fragmented, adding to the difficulties experienced by the working poor. But family can itself stifle human flourishing; the domestic ideology that then dominated mid-Victorian lives – and whose impact is felt by even the humblest within society, such as Blackpool, who cannot afford to divorce his wife and take up a relationship with his beloved Rachael – created a closed, and close-minded world, in which (gendered) roles were fixed. (As we have seen in the case of Louisa, this ideology placed particular constraints on well-to-do women; the less well-off, like Rachael, still worked, and often had no choice but to work). As social scientists emphasise, however, this kind of rigid, closed family dynamic creates its own crises, as does Louisa's over-determined trajectory. An open family, by contrast, adapts to change and changing environments: it is responsive and therefore resilient, in part because of the importance it attaches to constitutive relationships (Howe 2009, 110). In an open family, notes David Howe, people 'think *interpersonally*, not intrapsychically' (2009, 110). To do so is, in itself, an imaginative act – an act of 'Fancy', as it were – but it underpins a self-supporting community that might then be characterised as kind of nondualistic assemblage; characterised, that is, by a general absence of the divisive rifts that characterise society as a whole. Class, gender, and species do not divide this little community in the way that they do in Gradgrind's hierarchical and dualistic world.

The women of Sleary's company illustrate the novel's point. They appear more or less impervious to the demand that they be demure home-makers, and angels of their houses. They instead demonstrate what might then have seemed a most remarkable (shocking?) ability to do everything that the men do, and perhaps more:

> All the mothers could (and did) dance, upon the slack wire and the tight-rope, and perform rapid acts on bare-backed steeds; none of them were at all particular in respect of showing their legs; and one of them, alone in a Greek chariot, drove six in hand into every town they came to.
>
> (Bk. I, Ch. 6, 38)

Modest they are not; but then, there is nothing conventional about them. Even the arithmetic of husbands, wives, and offspring is left a little unclear.

As the narrator casually observes of the company, '[t]here were two or three handsome young women among them, with their two or three husbands, and their two or three mothers, and their eight or nine little children' (Bk. I, Ch. 6, 38). 'They all assumed to be mighty rakish and knowing', the narrator adds, 'they were not very tidy in their private dresses, they were not at all orderly in their domestic arrangements, and the combined literature of the whole company would have produced but a poor letter on any subject' (Bk. I, Ch. 6, 38). They are, in a word, bohemian, a term that conflates unconventionality and creativity (Baker 2018, 290), both of which position them at the very margins of what a factful society considers moral and acceptable. This too bothers them not at all. To the contrary, they seem to have arranged themselves in a way that brings out the best in themselves and each other. Their own lack of book-learning (Bk. I, Ch. 6, 38) – in Gradgrind's sense of 'system, schedule, blue book, report, and tabular statements A to Z' (Bk. I, Ch. 9, 57–8) – is no obstacle to the creation of a benign, tolerant, free-spirited and free-thinking community, in which the individual flourishes along with the collective. As the narrator points out,

> there was a remarkable gentleness and childishness about these people, a special inaptitude for any kind of sharp practice, and an untiring readiness to help and pity one another, deserving often of as much respect, and always of as much generous construction, as the every-day virtues of any class of people in the world.
>
> (Bk. I, Ch. 6, 38–9)

'[A]ny class of people in the world': Dickens quietly slips into his description of the circus folk a much more radical observation – that they share their goodness with the common run of people everywhere. The only constitutive difference between them is that, unlike the products of Gradgrind's educational system (such as Bitzer), the circus folk have escaped the impress of a philosophy of 'Fact', which would reduce society to an aggregation of self-interested (and by definition selfish) individuals, each competing with the other. Gradgrind has, in other words, made the worst of those whose lives he has shaped, not the best, which underlines the meaning of Sleary's own remark, twice pointedly addressed to Gradgrind: 'Do the withe thing and the kind thing too, and make the betht of uth; not the wurtht!' (Bk. III, Ch. 8, 269) '... I conthider that I lay down the philothophy of the thubject when I thay to you, Thquire, make the betht of uth: not the wurtht!' (Bk. I, Ch. 6, 43–4).

By the end of the novel, Gradgrind has good reason to appreciate Sleary's meaning. His own efforts to construct the model citizen have, in the case of son Tom, daughter Louisa, and sometime model pupil Bitzer, failed in spectacular but quite unexpected ways. By contrast, his inability to educate Sissy simply reflects her own goodness, her own relational ethic, as a product of the circus community. Fail he most certainly does. In one, notable scene, Sissy relates how, try as she might, she can never seem to find the right answers to her teacher's questions:

> after eight weeks of induction into the elements of Political Economy, she had only yesterday been set right by a prattler three feet high, for returning to the question, 'What is the first principle of this science?' the absurd answer, 'To do unto others as I would that they should do unto me'.
>
> (Bk. I, Ch. 7, 57)

The 'absurd answer' that Sissy supplies is a Biblical precept (Luke 6:31) widely known as the Golden Rule (Kirk 2003), sometimes (if problematically) regarded as a universal principle of reciprocity, 'a path towards altruism' (Fiala 2009, 24), and the basis of any society that aspires to be a moral, just, and fair one. Worse, Sissy makes the dreadful mistake of thinking that a nation with 'fifty millions of money' is not necessarily a prosperous one, depending on who has the millions, and creates further outrage when she wonders whether a city of a million people in which 'only five-and-twenty are starved to death in the streets' is self-evidently a good one, since it is just as hard on the starving whether the city's size is 'a million, or a million million' (Bk. I, Ch. 7, 59). Worst of all, she mistakes 'National Prosperity' for 'Natural Prosperity' (Bk. I, Ch. 7, 58), which are not, to her great surprise, the same thing, a 'mistake' that deftly highlights the way in which the political economy whose lessons she fails to grasp has itself failed to understand the ultimate source of all wealth. Thus, Sissy's apparently unthinking, heedless observations – which to her interlocutors seem to be so much nonsense – operate as subtle (or not so subtle) indictments of the society her social superiors are intent on making, with its skewed sense of what constitutes the common wealth. It is Sissy, by contrast, who appreciates the ties that bind – the entanglements that constitute happiness and meaning – and Sissy who is always 'trying hard to know her humbler fellow-creatures' (Bk. III, Ch. 9, 274); as Gradgrind belatedly realises, it is Sissy who has quietly sought to knit together and save the broken fragments of his family, 'like a good fairy in his house' (Bk. III, Ch. 7, 254–5).

Sissy's values are, of course, those of the circus in which she was raised. What, then, are we to make of Dickens's depiction of it, as society's alternate? Might it, perhaps, be described as a kind of utopia, as Michael Hollington does (2012, 136)? The word 'utopia' literally means a no place, and, appropriately, the circus is first encountered on the edge of town, where it occupies the 'neutral ground [that is] neither town nor country' (Bk. I, Ch. 3, 15). But if the circus is a utopia, it is a very curious one. Shorn of the idealised elements that we might expect from 'Fancy's' alternate to a self-interested society, the world of Dickens's circus is (as I have argued) a gritty, hard-working one. Moreover, and whilst the world of the utopian ideal is typically both carefully contrived *and* carefully controlled – like the one that Gradgrind is himself endeavouring to create – the circus appears disorganised, unsystematic, even anarchic. Perhaps the circus might, instead, be described as a larger, more extensive, and still more idiosyncratic version of the community which Esther Summerson establishes in *Bleak House*, as itself a family that spans the generations and encompasses friends as well as relatives. But whereas Esther's own Bleak House operates as a kind of neo-pastoral community – and as such, an extension of the 'rural homeliness' whose presentation George Gissing identified as one of Dickens's 'supreme merits' (2015, 172) – the circus is something else again: it is nomadic. The circus people are, as Mr Childers explains, never stationary, but 'comers and goers anywhere' (Bk. I, Ch. 6, 37). The circus is always on the move, like a wooden-walled ark in a sea of self-interested troubles (the circuses of the time were indeed constructed around wooden booths, not tents; Schlicke 1985 [1988], 168).

It may be argued, of course, that the nomadism of the circus is a symptom of its marginal status within society, a society on which it nevertheless depends to make a living. But it may also be argued that its nomadism is a source of strength: that the circus embodies what posthumanist scholars have defined as a nomadic sensibility, as 'an inherent understanding of the vagaries and contingencies of life on the move' (Baker 2018, 290), *and* (after the work of Rosi Braidotti) a nomadic subjectivity, adaptable, resilient, but also transgressive, where 'subjectivity includes relations to a multitude of nonhuman "others"', and '"life" is not only defined as *bios*, but also as a *zoe*-centred, nonhuman process' (Braidotti 2018, 340). On the one hand, therefore, the circus offers 'a new way of imagining community that privileges social connection over geographic belonging – and the consistent, territorial and thus humanist basis it depends on' (Baker 2018, 290). On the other, it creates a 'symbolic space' within which 'democracy, equality and mutual support prevail' (Hollington 2012, 136), and identity is itself fluid, adaptable, adaptive – a fluidity which is, as scholars such as Braidotti have stressed (2011, 25), singularly important for the way in which it erodes the fixed structures of meaning that have been used to define and constrain women. As Braidotti argues, 'the subject "woman" is not a monolithic essence, defined once and for all, but rather the site of multiple, complex, and potentially contradictory sets of experiences, determined by overlapping variables such as class, race, age, lifestyle, and sexual preference', requiring a 'culturally differentiated understanding of the subject in general and of the feminist subject in particular' (2011, 25).

From this perspective, the circus does indeed embody a kind of utopian ideal, as a nondualistic, relational assemblage in which equity extends across gender, class, and species. No less intriguingly, Dickens's deterritorialised, utopian alternate finds a way to flourish without recourse to the pastoral notion of an idyllic green refuge, the traditional but arguably illusory literary antidote to life in the corrupting city (Gifford 1999, 18–19). As a community which has no roots – which has no ancestral homeland after which to hanker – the circus in *Hard Times* sidesteps the (artificial) dichotomy of town and country, and perhaps even of culture and nature.[7] Once again, its nomadism is a source of strength, and a sign of its adaptability in a modern world; it is not simply a measure of its ability to survive, but to thrive, in a self-contained and sustainable way that itself enacts (even if on the smallest of scales) a justice beyond self, beyond gender, and beyond species.

*

Industrial modernity contains many contradictions, but the most profound of them is that the accumulation it fosters 'is not only productive; it is necrotic', everywhere unfolding a narrative of 'slow violence' that 'devours all life' (McBrien 2016, 116). As Dickens observes in *Hard Times*, 'progress' obscures itself with '[a] blur of soot and smoke ... murkily creeping along the earth' (Bk. II, Ch. 1, 105). This sepulchre-like gloom denotes the emergence of what has been termed a Neocrocene, an emerging age of 'accumulation by extinction' (qtd in Moore 2016, 8) whose outlines Dickens saw emerging in and around the many Coketowns that industrial modernity was generating.

As Dickens realised, and depicted so memorably in the novel, the causes of society's great transformation lay not only in the free play of capitalism and unfettered industrial growth, but in a working out of the Enlightenment's rationalistic, modernising narrative, with its confident promise of progress. As Horkheimer and Adorno would later argue, a 'fully enlightened earth radiates disaster triumphant' (1997 [2010], 3). At once mechanistic, individualistic and reductionistic, Enlightenment rationality suggests that entities can be explained independently of their relationship with others and with their environment; in Dickens's fictive world, by contrast, his subjects are never separable from the constitutive relationalities on which they depend, relationalities that encompass the nonhuman and more-than-human, biotic and abiotic, natural and built world. A reified 'Nature' has no place in this argument: environment is a totality – an 'immersive phenomena' – that encompasses multiple interdependencies and interconnections, creating a 'mesh' in which Dickens's characters find themselves entangled (Morton 2010, 61). Everything is connected; everything, situated; nothing defies 'overall meaning' (Beer 1985, 46). Thus, Dickens's network of relationalities exceeds the definitional boundaries that mark out town and country, and even people and place.

By rejecting Enlightenment rationality, and in his concern for the actual wants and needs of real beings, Dickens also opposes any interpretation of human behaviour which is narrowly mechanistic or materialistic. Humans are not machines, to be programmed as such. Nor are their needs simply material ones. Industrial workers might well demand change, but not (or not simply) because they are hungry. He has in mind a larger understanding of what constitutes a flourishing existence. As the Leavises argued, Dickens is 'unmistakably possessed by a comprehensive vision' (1970, 188), and in

Hard Times, he offers nothing less than '[t]he confutation of Utilitarianism by life' (196) – by 'generous, impulsive life' (191).

This sense of life at its fullest finds its most powerful expression in Dickens's depiction of the circus community, a community which embodies both a spatial, nomadic sensibility, and the 'multiple belongings' of a zoe-centred nomadic subjectivity (Braidotti 2011, 10; 2018, 340). As Dickens emphasises, and makes clear through his depiction of Sissy, that community embraces a relational ethic of care that transgresses class, gender, and species boundaries, and generates its own, idiosyncratic form of emergent justice. As Dickens stresses, that emerging awareness of self in relation to others – to all those society has 'othered' – requires care, compassion, a sense of the constitutive connections that encompass all forms of life and the environs that sustain and surrounds life; it requires empathy and understanding, as Sissy so notably demonstrates. Above all, it requires 'Fancy', the term that Dickens uses to denote the imagination. In a narrowly self-interested world defined by 'Fact', argues Dickens, it is 'Fancy' that enables us to see beyond the self, and it is a fanciful approach which enables *Hard Times* to exceed the constraints of realism itself, with its emphasis on the subject as inviolable and sovereign, separate from its own constitutive entanglements with others.

To this extent, and whilst the word 'ecology' had yet to be coined and 'environmentalism' had yet to emerge as a pressing and distinct concern, Dickens may fairly be described as 'an ecologically minded novelist' (Boone 2017, 106). But the circus is not quite the sum of Dickens's argument, notable as it is for its role as Coketown's antithesis and antidote. Dickens's depictions of lively, agential realities – and of the complex, constitutive entanglements that make up all forms of existence – position him not only as a proto-ecological thinker, but as new materialist one (Parham 2017 [2019], 120–1). To the new materialist thinkers that his work anticipates, there is only a naturalcultural reality (Barad 2007, 32–4). Indeed, Morton argues that the very notion of a separate and distinct 'nature' is problematic (Morton 2007 [2009], 1–2); it signifies the existence of a pervasive divide that enables political economies like Gradgrind's to exploit as externalities the natural world with which human existence is always already entangled. It is this conceptual divide which the circus transcends, enabling it, instead, to focus on the myriad intra-actions that together make up a flourishing existence.

Yet it is precisely these intra-actions – at once performative, processual, constitutive – that point to the final aspect of Dickens's engagement with and response to environmental injustice: a 'properly ecological' society may well be one without 'nature' (Morton 2007 [2009], 1), but it is also one that fully recognises, acknowledges, and respects its intimate entanglement with the nonhuman and natural processes on which its flourishing depends. This is not just a question of breathable air, fertile soil, fresh water, or liveable climates, which we might consider the constitutive basics of life in nearly all its forms; there is another, subtler, psychological dimension to the needs that make up a truly flourishing existence, as Dickens also suggests. Can Coketown's 'blur of soot and smoke' (Bk. II, Ch. 1, 105) ever be escaped? Is there a world beyond 'the soot-prints of modernisation' (Taylor 2016, 47)? When Rachael and Sissy help themselves 'out of the smoke' one Sunday in autumn, thirsting for a 'draught of pure air' in the countryside (Bk. III, Ch. 6, 245), they are responding to a deep-rooted need for a world beyond their world, in which the nonhuman and natural have not been wholly banished or completely distorted by the impact of industrial modernity. They are responding, in other words, to 'the felt presence of an absence', and seeking a 'glimpse of an open whole' (Taylor 2016, 6) without which they are themselves incomplete. It may be that Coketown still marks the horizon with 'a black mist'; it may be that railroads crisscross the countryside, and mines mark its surface; but '[u]nder their feet, the grass was fresh; beautiful shadows of branches flickered upon it, and speckled it; hedgerows were luxuriant; everything was at peace' (Bk. III, Ch. 6, 245).[8]

What Dickens suggests, therefore, is that there is a continuing need for contact with these natural processes; that precisely because reality is at once natural and cultural – properly, naturalcultural, after the work of Karen Barad – human flourishing depends on these constitutive contacts. Their denial is properly a loss, and it impacts on human flourishing.[9] This was not, however, Dickens's insight alone. Dickens owed much of his critical inspiration to Thomas Carlyle; so did John Ruskin, Dickens's near contemporary. Famously, Ruskin championed an artistic (re)turn to 'nature', but over time, his art criticism gave way to social critique, as increasingly he recognised that the kind of art a society produced was itself dependent on its wider relationship to the nonhuman and natural world – a relationship that was everywhere being compromised by the capitalist drive for profit encapsulated in the doctrines of political economy. As Ruskin realised, and the next chapter discusses in more detail, society was increasingly at fault in what it chose to value – and discount – and those choices generated both social inequality and environmental crisis.

*

Notes

1 Thus, Karen Bell writes that '"[e]nvironmental classism" refers to policies or practices that impact less favourably on working-class individuals and groups with respect to the quality of their living, working and leisure environments' (2020, 3). Bell makes the point that, in contemporary environmental justice scholarship, questions of class have tended to take second place to the role of low incomes or race in shaping the nature of environmental injustice (2020, 5); in Britain, she contends, class remains a decisive factor.

2 The copy text for this chapter is the Oxford University Press edition (2006 [2008]) of Dickens's *Hard Times*, to which book, chapter, and page numbers refer.

3 *Hard Times* continues to generate ecocritical interest, not least Miller's own analysis (2021, 63–70) of the novel's 'broader provincial landscape of extraction' (63), a 'country of the old pits' already exhausted by industry's demands (Bk. III, Ch. 6, 246). Readers are also referred to Dewey W. Hall's analysis of 'factory schools and factory system' (2017, 52), in which he argues that figures such as Sissy are themselves 'signifiers of what may be called early environmental injustices' (52). In an alternative reading of Dickens's ecocritical significance, Troy Boone suggests that the status of novels such as *Hard Times* (exemplum of Dickens's later concerns as a social novelist) tends to occlude Dickens's interest in 'nonhuman natural realms' (2017, 98).

4 A capabilities approach sets out 'minimum core social entitlements' for human functioning, understood not just in terms of resources, but opportunities to act, to be (Nussbaum 2006, 75). As elaborated by Nussbaum, many of these capabilities fit readily with the demands of environmental justice activists, who also insist on the right to bodily health and bodily integrity (capabilities 2, 3; Nussbaum 2006, 76), and on 'control over one's environment', both political and material (capability 10; Nussbaum 2006, 77). Nussbaum also underlines the need for play (capability 9; 77), for a full emotional life (capability 5), and insists on the right 'to use the senses, to imagine, think, and reason' (capability 4; 76) – 'to form a conception of the good' (capability 6; 77) – and 'to live with and towards others' (capability 7; 77).

5 The idle younger son of an established family, Harthouse has been directed towards a career with the Gradgrind party (Bk. II, Ch. 2, 118). His only real interest lies in ingratiating himself into the company of Bounderby's new wife, Louisa, né Gradgrind. As Harthouse admits with a frankness intended to disarm, he is a 'sordid piece of human nature … and altogether incapable of any Arcadian proceedings whatever' (Bk. II, Ch. 7, 159).

6 Significantly, Nussbaum also relates her point to utilitarian thinking. In the life of western societies, she argues, a 'refusal of sympathy' is all too often 'aided and abetted' by 'technical ways of modelling human behaviour, especially those that derive from economic utilitarianism' (1995, xiii). It is these ways of rationalising human behaviour, and reducing it to questions of what is useful, that literature subverts, as Gradgrind recognises (Nussbaum 1995, 1–2).

7 As Tristan Sipley points out, the critical tendency to focus 'either on the "pastoral Dickens" or the "gritty urban Dickens"' risks reinforcing an artificial 'epistemological polarity', when the 'value of Dickens lies precisely in the way his sprawling fictions avoid fixating on any space in isolation' (2011, 18).

8 Perhaps this is an absence that the circus community (human and nonhuman) itself feels; as it makes its way from one piece of dead ground (or edgeland) to another, its nomadism also marks out a kind of ecological exile, for them no less than for those denizens of Coketown, forced to live amidst environmental degradation and damage (Gladwin 2018, 5). Dickens does not elaborate. Nor does Dickens develop the theme in relation to Sissy and Rachel, and their Sunday afternoon pilgrimage. Having found a place beyond the smoke, they immediately discover a clue to the fate of Stephen Blackpool (Bk. III, Ch. 6, 246), and their intra-action with an agential, nonhuman reality is forgotten, as the plot takes its next stride forward. It may be, perhaps, that Dickens has no real interest in the nonhuman and natural, that his lyrical effusions about 'Nature' are little more than simply clichéd gestures, a view that Troy Boone sets out to contradict (2017, 97–101). A more compelling response is that Dickens does not need to develop his theme: his readers, as themselves newcomers to the abnatural world of the nation's new Coketowns, would have understood his meaning without lengthy explanation. Although the census of 1851 showed that Britain was now an urban

society – the first in world history – many of those who lived in the towns and cities had migrated to them from the country: 'in 1851, of every 33 people who lived in a city, only 13 had been born there' (Steinbach 2017, 14). Thus, most of Dickens's readers were familiar with the natural, nonhuman processes to which he refers, and understood why Sissy and Rachael felt the need to make their pilgrimage: their formative experience and collective memories supply what Dickens does not, exerting a 'tacit pressure on the narrative' (Chase and Levenson 2009, 132).

9 As a growing body of evidence underlines (see, for example, Baxter and Pelletier, 2019), human nature seeks out and needs these kinds of contact; as Nussbaum argues, these contacts form a part of what constitutes human flourishing (Nussbaum 2006, 77). The need may be innate, instinctual, since 'everything human [argues Harold Fromm] derives from the evolved body and brain' (2009, 7). This is not to ignore the cultural, constructed dimension of human consciousness, simply to emphasise (as do new materialist thinkers) the powerful influence of our own materiality, and the depth of its entanglement with the environment from which it is never finally separable. Seen from this perspective, contact with the nonhuman, natural world may well be a foundational part of human life itself, and not in any way ancillary to it. These points notwithstanding, the contrary argument may also be made: that, as Simon Estok argues, humans do not possess an innate biophilia, but an innate ecophobia; we are, Estok contends, genetically encoded with an 'antipathy towards nature' that once helped 'preserve our species' (Estok 2018, 1).

Works cited

Adorno, Theodor W., and Max Horkheimer. 1997 [2010]. *Dialectic of Enlightenment*, translated by John Cumming. London: Verso. First published 1944.

Altvater, Elmar. 2016. 'The Capitalocene, or, Geoengineering against Capitalism's Planetary Boundaries'. *Anthropocene of Capitalocene? Nature, History, and the Crisis of Capitalism*, edited by Jason W. Moore, 138–52. Oakland, CA: PM Press.

Baker, Daniel. 2018. 'Nomadic Sensibility'. *Posthuman Glossary*, edited by Rosi Braidotti and Maria Hlavajova, 290–2. London: Bloomsbury Academic.

Barad, Karen. 2007. *Meeting the Universe Halfway: Quantum Physics and the Entanglement of Matter and Meaning*. London: Duke University Press.

Barker, Hannah, and Elaine Chalus. 2005. 'Introduction'. *Women's History: Britain*, 1700–1850, edited by Hannah Barker and Elaine Chalus, 1–8. Abingdon: Routledge.

Baxter, D. E., and L. G. Pelletier 2019. 'Is Nature Relatedness a Basic Human Psychological Need? A Critical Examination of the Extant Literature'. *Canadian Psychology/Psychologie Canadienne* 60 (1): 21–34. doi:10.1037/cap0000145

Beck, Ulrich. 1992 [2012]. *Risk Society: Towards a New Modernity*, translated by Mark Ritter. London: Sage.

Beer, Gillian. 1985. *Darwin's Plots: Evolutionary Narrative in Darwin, George Eliot and Nineteenth-Century Fiction*. London: Ark.

Bell, Karen. 2020. *Working-Class Environmentalism: An Agenda for a Just and Fair Transition to Sustainability*. Cham, Switzerland: Palgrave Macmillan.

Berger, John. 2009. *Why Look at Animals?* London: Penguin.

Boone, Troy. 2017. 'Early Dickens and Ecocriticism: The Social Novelist and the Nonhuman'. *Victorian Writers and the Environment*, edited by Laurence W. Mazzeno and Ronald D. Morrison, 97–113. London: Routledge.

Braidotti, Rosi. 2018. 'Posthuman Critical Theory'. *Posthuman Glossary*, edited by Rosi Braidotti and Maria Hlavajova, 339–42. London: Bloomsbury Academic.

Braidotti, Rosi. 2011. *Nomadic Subjects: Embodiment and Sexual Difference in Contemporary Feminist Theory*. 2nd edn. New York, NY: Columbia University Press.

Carlyle, Thomas. 1843 [1897]. *Past and Present. The Works of Thomas Carlyle, Volume Six*, 417–500. New York, NY: Peter Fenelon Collier. Available at archive.org, https://archive.org/details/worksofthomascar0006carl/page/n5/mode/2up?view=theater. Accessed 15 December 2022.

Carlyle, Thomas. 1829 [1897]. 'Signs of the Times'. *The Works of Thomas Carlyle, Volume Fourteen*, 462–87. New York, NY: Peter Fenelon Collier. Available at archive.org, https://archive.org/details/worksthomascarl07carlgoog/page/n12/mode/2up?view=theater. Accessed 14 December 2022.

Chase, Karen, and Michael Levenson. 2009. 'Green Dickens'. Contemporary Dickens, edited by Eileen Gillooly and Deirdre David, 131–51. Columbus, OH: Ohio State University Press.

Chinn, Carl. 1995 [2006]. *Poverty amidst Prosperity: The Urban Poor in England, 1834–1914*. Lancaster: Carnegie.

Clapp, B. W. 1994. *An Environmental History of Britain since the Industrial Revolution*. London: Longman.

Cook, E. T., and Alexander Wedderburn (eds). 1903–12. *The Library Edition of John Ruskin's Works*. 39 vols. London: George Allen. Available at www.lancaster.ac.uk/the-ruskin/the-complete-works-of-ruskin/. Accessed 5 January 2023.

Daunton, Martin. 2000. 'Society and economic life'. *The Nineteenth Century*, edited by Colin Matthew, 41–82. Oxford: Oxford University Press.

Derrida, Jacques. 2008. *The Animal That Therefore I Am*, translated by David Wills, edited by Marie-Louise Mallet. New York, NY: Fordham University Press.

Dickens, Charles. 1853. 'Frauds on Fairies'. *Household Words, Volume 8*, 1 October: 97–110. Available at Dickens Journal Online, www.djo.org.uk/household-words/volume-viii/page-97.html. Accessed 10 September 2022.

Dickens, Charles. 2003. *Bleak House*, edited and introduced by Nicola Bradbury, preface by Terry Eagleton. London: Penguin.

Dickens, Charles. 2006 [2008]. *Hard Times*, edited and introduced by Paul Schlicke. Oxford: Oxford University Press.

Estok, Simon C. 2018. *The Ecophobia Hypothesis*. London: Routledge.

Fiala, Andrew. 2009. '"How Would You Like to Be Him?" The Golden Rule, Third Person Descriptions, and Virtue Ethics'. *The Pluralist* 4 (2): 24–37. Available at jstor.org, www.jstor.org/stable/20708975. Accessed 2 September 2022.

Forster, John. Not dated. *The Life of Charles Dickens*. London: Chapman & Hall. First published 1872–4.

Fromm, Harold. 2009. *The Nature of Being Human: From Environmentalism to Consciousness*. Baltimore, MD: John Hopkins University Press.

Gaskell, Elizabeth. 2006 [2008]. *Mary Barton: A Tale of Manchester Life*, edited by Shirley Foster. Oxford: Oxford University Press.

Gifford, Terry. 1999. *Pastoral*. London: Routledge.

Gissing, George. 2015. *Collected Essays*, edited by Pierre Coustillas. Surrey: Grayswood.

Gladwin, Derek. 2018. *Ecological Exile: Spatial Injustice and Environmental Humanities*. London: Routledge.

Hall, Dewey W. 2017. 'Hard Times: Factory Education, Factory System, and the Preston Strike'. *Victorian Ecocriticism: The Politics of Place and Early Environmental Justice*, 51–68. London: Lexington.

Hansard HC [House of Commons] Debate 129 (1495–7), 8 August 1853. Available from hansard.parliament.uk, https://hansard.parliament.uk/Commons/1853-08-08/debates/a4437a20-326e-4b19-a4ab-232fe434221e/SmokeNuisanceAbatement(Metropolis)Bill. Accessed 16 December 2022.

Harvey, Karen. 2005. 'Sexuality and the Body'. *Women's History: Britain*, 1700–1850, edited by Hannah Barker and Elaine Chalus, 78–99. Abingdon: Routledge.

Holifield, Ryan, Jayajit Chakraborty, and Gordon Walker. 2018. 'Introduction: The Worlds of Environmental Justice'. *The Routledge Handbook of Environmental Justice*, edited by Ryan Holifield, Jayajit Chakraborty, and Gordon Walker, 1–12. Abingdon: Routledge.

Hollington, Michael. 2012. 'Dickens and the Circus of Modernity'. *Dickens and Modernity*, edited by Juliet John, 133–49. Cambridge: D. S. Brewer.

Howe, David. 2009. *A Brief Introduction to Social Work Theory*. Basingstoke: Palgrave Macmillan.

Kirk, Alan. 2003. "Love Your Enemies', the Golden Rule, and Ancient Reciprocity (Luke 6: 27–35)'. *Journal of Biblical Literature* 122 (4): 667–86. doi:10.2307/3268071

Leavis, F. R., and Q. D. Leavis. 1970. *Dickens the Novelist*. London: Chatto & Windus.

Levine, George. 1981 [1983]. *The Realistic Imagination: English Fiction from Frankenstein to Lady Chatterley*. Chicago, IL: University of Chicago Press.

Luckin, Bill. 2015. *Death and Survival in Urban Britain: Disease, Pollution and Environment 1800–1950*. London: Bloomsbury Academic.

Matthew, Colin. 2000. 'The United Kingdom and the Victorian Century, 1815–1901'. *The Nineteenth Century*, edited by Colin Matthew, 1–38. Oxford: Oxford University Press.

McBrien, Justin. 2016. 'Accumulating Extinction: Planetary Catastrophism in the Neocrocene'. *Anthropocene of Capitalocene? Nature, History, and the Crisis of Capitalism*, edited by Jason W. Moore, 116–37. Oakland, CA: PM Press.

Miller, Elizabeth Carolyn. 2021. *Extraction Ecologies and the Literature of the Long Exhaustion*. Princeton, NJ: Princeton University Press.

Moore, Jason W. 2016. 'Introduction'. *Anthropocene of Capitalocene? Nature, History, and the Crisis of Capitalism*, edited by Jason W. Moore, 1–11. Oakland, CA: PM Press.

Morton, Timothy. 2007 [2009]. *Ecology Without Nature: Rethinking Environmental Aesthetics*. London: Harvard University Press.

Morton, Timothy. 2010. *The Ecological Thought*. London: Harvard University Press.

Morton, Timothy. 2013. *Hyperobjects: Philosophy and Ecology after the End of the World*. London: University of Minnesota Press.

Nixon, Rob. 2011. *Slow Violence and the Environmentalism of the Poor*. Cambridge, MA: Harvard University Press.

Nussbaum, Martha C. 1995. *Poetic Justice: The Literary Imagination and Public Life*. Boston, MA: Beacon Press.

Nussbaum, Martha C. 2003. 'Capabilities as Fundamental Entitlements: Sen and Social Justice'. *Feminist Economics* 9 (2–3), 33–59. doi: org/10.1080/1354570022000077926

Nussbaum, Martha C. 2006. *Frontiers of Justice: Disability, Nationalities, Species Membership*. Cambridge, MA: Belknap Press.

Parham John., 2017 [2019]. 'Bleak Intra-Actions: Dickens, Turbulence, Material Ecology'. *Victorian Writers and the Environment: Ecocritical Perspectives*, edited by Laurence W. Mazzeno and Ronald D. Morrison, 114–29. Abingdon: Routledge.

Plumwood, Val. 1993 [1997]. *Feminism and the Mastery of Nature*. London: Routledge.

Poovey, Mary. 1995. *Making a Social Body: British Cultural Formation, 1830–1864*. London: University of Chicago Press.

Posner, Richard A. 1997. 'Against Ethical Criticism.' *Philosophy and Literature* 21, 1–27.

Schlicke, Michael. 1985 [1988]. *Dickens and Popular Entertainment*. London: Unwin Hyman.

Sen, Amartya. 2006 [2007]. *Identity and Violence: The Illusion of Destiny*. London: Penguin.

Sipley, Tristan. 2011. 'The Revenge of 'Swamp Thing': Wetlands, Industrial Capitalism, and the Ecological Contradiction of *Great Expectations*'. *Journal of Ecocriticism* 3 (1): 17–28. Available at https://ojs.unbc.ca/index.php/joe/article/view/143. Accessed 14 January 2023.

Slater, Michael. 2011. *The Genius of Dickens*. London: Duckworth Overlook.

Stein, Rachel. 2004. 'Introduction'. *New Perspectives on Environmental Justice: Gender, Sexuality, and Activism*, edited by Rachel Stein, 1–20. Piscataway, NJ: Rutgers University Press.

Steinbach, Susie L. 2017. *Understanding the Victorians: Politics, Culture and Society in Nineteenth-Century Britain*. 2nd edn. Abingdon: Routledge.

Stevenson, John. 1993. 'Social aspects of the Industrial Revolution'. *The Industrial Revolution and British Society*, edited by Patrick K. O'Brien and Roland Quinault, 229–53. Cambridge: Cambridge University Press.

Taylor, Jesse Oak. 2016. *The Sky of Our Manufacture: The London Fog in British Fiction from Dickens to Woolf*. London: University of Virginia Press.

Voth, Hans-Joachim. 2004. 'Living Standards and the Urban Environment'. *The Cambridge Economic History of Modern Britain, Volume 1: Industrialisation*, 1700–1860, edited by Roderick Floud and Paul Johnson, 268–94. Cambridge: Cambridge University Press.

Wilmer, Clive. 1985. 'Introduction'. *John Ruskin, Unto This Last and Other Writings*, edited and introduced by Clive Wilmer, 7–37. Harmondsworth: Penguin.

Žižek, Slavoj. 2008 [2009]. *Violence: Six Sideways Reflections*. London: Profile.

6 John Ruskin's *Unto this Last*: towards a 'deeper felicity'

Born in 1819, the great Victorian polymath John Ruskin was celebrated and later derided in his own lifetime, and largely neglected thereafter (Rose 2003, 87). Today, the majority of his work is out of print.[1] This is perhaps not surprising: his works are voluminous, his prose difficult, and his meaning sometimes elusive or self-contradictory (Eagles 2007, 84–5). When his meaning is clear, it is often problematic; his views on the 'separate spheres' of men and women today astonish, as Phyllis Rose points out (2003, 91). Yet Ruskin remains one of the most fascinating and insightful Victorian commentators on the changing nature of modern society, and its relationship to its environs. Not surprisingly, ecocritical interest in his work has grown steadily over recent years.[2] That interest reflects the many ways in which his concern with 'the whole system of nature' (Cook and Wedderburn 1903–12, Vol. III, 616)[3] anticipated the emerging study of ecology (see, for example, Frost 2017 [2019], 13–15; MacDuffie 2014, 137–69; Williams 2016, 319–20). From the perspective of a material ecocriticism, however, Ruskin's work is particularly fascinating because he did not confine his interest *to* the natural world. Strongly as he felt the influence of his Romantic inheritance (Shionoya 2014, 16), Ruskin took as great an interest in the built world as in the living, natural one. No less significantly, Ruskin related his interest in the material world to the society that was (re)shaping it. Influenced in part by the example of Thomas Carlyle's own social criticism (Shionoya 2014, 18), Ruskin's concern with art and architecture developed into an increasingly sceptical interrogation of the society which produced it. Great art and architecture were the product of a healthy society, Ruskin contended, and a flourishing, natural world was an integral part of it (Eagles 2007, 85); but as he insisted, such a thing was impossible in a society possessed by the 'darkening, denaturing, and immiserating force' of industrial modernity (Williams 2016, 319), and dominated by the false religion of 'Mammon', which comprehensively failed to understand what was valuable. There was much more to wealth, Ruskin insisted, than the 'myopic view' of it 'as profit and frivolous consumption' (Williams 2016, 321). Anticipating much more recent critical responses to the capitalist dominance of humanity-in-nature, Ruskin argued 'not only for a more equitable distribution of wealth', but for what we would now regard as 'a *new conception of wealth*, in which equity and sustainability in the

DOI: 10.4324/9780367821609-7

reproduction of life (of *all* life) is central to our vision of the future' (italics in the original; Moore 2016, 10).

These ideas formed the basis of the four essays Ruskin published in 1860, which have together become known as *Unto this Last*. Notwithstanding the importance of later works such as his addresses to workmen, *Fors Clavigera* (1871–84), and his lectures on *The Storm-Cloud of the Nineteenth Century* (1884), these essays are an eloquent and uncompromising statement of his developing views: they mark his own, wide-ranging response to the convergence of 'pollution and poverty' (Williams 2016, 320) that beset a nation now structurally transformed by industrial capitalism at its most anarchic (Hilton 2006, 578). It was this unfettered form of capitalism on which the supposed 'science' of political economy insisted, and it was that science which Ruskin in turn criticised (he reserved particular opprobrium for John Stuart Mill's defence of its principles, although as I also discuss in this chapter, Mill's work itself contains a surprising thought experiment in ecological awareness). In his own response to the problems engendered by political economy, Ruskin argued for a broader understanding of wealth that took account of all those constitutive entanglements on which a flourishing society depended, and for an 'eco-systemic awareness' (MacDuffie 2014, 167) that extended to and encompassed the nonhuman and natural world. This is not to suggest that Ruskin acknowledged an equality – an ethical equivalence – between human and nonhuman, but by insisting on the need for a deeper understanding of what constituted the well-being or felicity of society, he also acknowledged the importance of a sense of justice that encompassed 'wider forms of reciprocity' (Birch 2012, 33).

*

Ruskin's critique of a modernising but profoundly unequal society begins and ends with the need to see things as they are (Levine 2000, 75). 'To see clearly', he declared in Volume III of *Modern Painters* (1856), 'is poetry, prophecy, and religion, – all in one' (V, 333); '[f]alse seeing is unseeing' (VII, 211). Sight mattered, because 'to see is to gain word and thought at once, and both true' (XVI, 180). As Ruskin insisted, the act of seeing clearly required 'serious and significant work' his was, as Caroline Levine observes, 'a *labouring aesthetic*' (italics in the original; 2000, 75) – but this way lay true enlightenment (XVI, 180). If we could but see the facts, Ruskin declared in his Inaugural Address at the Cambridge School of Art (1858), 'we should soon make it a different world' (XVI, 180). In his early works, however, Ruskin was concerned with how best to advance the cause of art and architecture, not social or environmental justice. If he insisted on the importance of seeing things clearly, it was because great art was simply impossible without first acquiring that skill. '[T]ruth to nature' demanded it (III, 617).

Nature, then, was Ruskin's touchstone, but in his reading of it he developed two concepts that were more generally applicable to his work and thought, and in turn relevant to his developing sense of what constituted social and environmental justice: as Mark Frost explains, Ruskin called for his readers to take note of 'what he termed the "Vital Beauty" and "leading lines" of nature' (2017 [2019], 15). Whereas 'Typical Beauty' was the concomitant of Ruskin's evangelical desire to trace God's handiwork in nature, 'Vital Beauty' characterised the 'felicitous [meaning appropriate, apposite, or advantageous] fulfilment of function in living things' (IV, 64) (see Shionoya 2014, 23). Vital Beauty was the dynamic, active corollary to those qualities of beauty 'constant in their address to human nature' (IV, 62), and in itself important, and important precisely because it encompassed all living things, and not just humankind (IV, 64). In Ruskin's mind's eye, notes Clive Wilmer, '[d]ifference is not imperfection but multitudinousness, variety, redundance, indeed wealth', each thing in its very particularity a 'unique manifestation' of God's love (1996, 90). This did not rule out a hierarchical reading of nature that positioned man as preeminent; as Frost adds (2017 [2019], 16), the right reading of nature underlined Ruskin's view 'that we should not only love all creatures well, but esteem them in that order which is according to God's laws' (IV, 161). Nevertheless, that reading did acknowledge that the right to a flourishing life – to 'the joyful and right exertion of perfect life' (IV, 64) – was not confined to humankind. It also followed, Ruskin told his readers in Volume II of *Modern Painters* (1846), that to esteem the nonhuman other was also to let it alone, never

> interfering with the working of nature in any way; nor, when we interfere to obtain service, [striving after] new forms such as nature never intended [or] substituting for the true and balanced beauty of the free creature some morbid development of a single power.
>
> (IV, 161)[4]

Ruskin's respect for the agential and autopoetic in nature – and his demand that nature be left to its own self-creation, without human interference – carries over into his later lecture on 'The Work of Iron' (1858) (see Frost 2017 [2019], 22–3). It was perfectly possible, Ruskin told his audience, to put iron to work in its pure form, as 'pokers, and pans', 'swords and scissors' (XVI, 377), but once oxidised, was it not more useful still? Iron might seem to be inanimate and dead, but the process of rusting demonstrated that this was not so: combining with the air, iron became living, dynamic, and part of processes on which human existence itself depended. By breathing, Ruskin declared, iron makes 'the ground we feed from, and nearly all the substances first needful to our existence' (XVI, 377). Ruskin's point is not simply that iron is more useful in this state than in another, or that it thereby supports all organic life; this 'vital change' transforms it into a living thing, with which our own existence is intimately entangled. 'Is it not strange', he asked, 'to find this stern and strong metal mingled so delicately in our human life that we cannot even blush without its help?' (XVI, 384).

As Frost underlines, 'The Work of Iron' reflects Ruskin's already highly advanced understanding of ecological interdependence (2017 [2019], 23); of matter as active, and the body as trans-corporeal, 'always intermeshed with the more-than-human world' (Alaimo 2010, 2). It also reflects his own, growing anxiety over industrial modernity's power to transform that world, and in so doing, materially affect and impoverish the lives of all those dependent on it. Increasingly, the earth was regarded as dead matter to be exploited without limit or consequence. '[I]ron rusted is Living', Ruskin told his audience; 'but when pure or polished, Dead' (XVI, 376–7). Which, he inquired, would they rather have?

> You think, perhaps, that your iron is wonderfully useful in a pure form, but how would you like the world, if all your meadows, instead of grass, grew nothing but iron wire – if all your arable ground, instead of being made of sand and clay, were suddenly turned into flat surfaces of steel – if the whole earth, instead of its green and glowing sphere, rich with forest and flower, showed nothing but the image of the vast furnace of a ghastly engine – a globe of black, lifeless, excoriated metal?
>
> (XVI, 378)

It is an arresting and bleak vision – a vision of 'a landscape literally disfigured by so-called industry' (Eagles 2007, 85) – and with language like this, Ruskin hoped to expose the short-sightedness of an exploitative and exceptionalist logic, with its assumption of humankind's hyper-separation from and superiority to 'nature' (Plumwood 1993 [1997], 48–55). To render the world lifeless would, after all, be to put an inevitable end to all forms of life, human life included.

How did Ruskin come to this conclusion? Ruskin's concept of 'Vital Beauty' emphasised the importance of 'felicitous fulfilment of function' (above, IV, 64), and provided the ethical grounding for his understanding of a flourishing existence, but the concept of 'leading lines' provided him with the epistemological means of understanding what that existence might look like. In *The Elements of Drawing* (1857), Ruskin explained 'the importance of leading or governing lines' to an artist's understanding (and rendering) of his subject (XV, 91). These lines, wrote Ruskin, were 'always expressive of the past and present action of the thing' (XV, 91). Tracing these lines enabled the artist to read a thing as text; to experience its history, understand its present, and identify the myriad interactions – or, after Karen Barad, 'intra-actions' (2007, 33) – on which it depended for its existence. This was the 'vital truth' of an entity, and for an artist, it was the key to rendering 'every natural form' (XV, 91) in all its changefulness and changeability:

> the animal in its motion, the tree in its growth, the cloud in its course, the mountain in its wearing away. Try always, whenever you look at a form,

> to see the lines in it which have had power over its past fate and will have power over its futurity.
>
> (XV, 91)

Thus, Ruskin's 'reading process involved tracing stories that stressed interconnectedness and dynamic processes' (Frost 2017 [2019], 18). Crucially, there was for Ruskin a further dimension to this 'interconnected web' (Coit 2010, 215): the way in which these interdependent parts worked together, as a whole, a view Ruskin derived from his spiritual belief. '[T]he natural phenomena under whose influence we exist', Ruskin later wrote, are 'the work and the gift of a Living Spirit greater than our own' (XXVI, 334). For Ruskin, operating under the influence of the evangelical faith in which he had been raised (Birch 2012, 32), it was impossible to see the natural world except as the expression of divine will, and as such, part of a beneficent scheme whose law was cooperation, not competition.[5] In the final volume of *Modern Painters* (1860), he set out what he called 'The Law of Help'. 'A pure or holy state of anything', he wrote, 'is that in which all its parts are helpful or consistent' (VII, 207):

> The highest and first law of the universe – and the other name of life is, therefore, 'help'. The other name of death is 'separation'. Government and co-operation are in all things and eternally the laws of life. Anarchy and competition, eternally, and in all things, the laws of death.
>
> (VII, 207)

On the one hand, Ruskin's belief in a Law of Help predisposed him to reject any reading of nature and of nature's processes which pointed, instead, to competition as a directing principle. On the other, it led him to reject the direction that contemporary society was taking: increasingly, his own '[e]xuberant observation of nature is supplanted by a sober consideration of modern self-interest' (Birch 2004 [2009], xiv), a self-absorbed and competitive commitment to 'progress' that was itself a threat to the nonhuman and natural world. Nature and society were interlinked, co-dependent and co-constitutive; now, a myopic, complacent, and self-interested society was destroying the very beauty that Ruskin had sought to champion. 'Once I could speak joyfully about beautiful things', he wrote; 'now I cannot anymore; for it seems to me that no one regards them. Wherever I look or travel in England or abroad, I see that men, wherever they can reach, destroy all beauty' (VII, 422–3).

In Ruskin's view, however, beauty was not confined to the natural world. The same emphasis on leading lines that enabled Ruskin to trace the emergence of mountain, tree, wave or cloud (XV, 91) also made him mindful of the stories that buildings themselves told; of time's passing, and the 'actual beauty in the marks of it' (VIII, 235). As he wrote in *The Seven Lamps of*

Architecture (1849), 'the greatest glory of a building is not in its stones, nor in its gold'.

> Its glory is in its Age, and in that deep sense of voicefulness, of stern watching, of mysterious sympathy, nay, even of approval or condemnation, which we feel in walls that have long been washed by the passing waves of humanity. It is in their lasting witness against men …
>
> (VIII, 233)

It was in and through 'that golden stain of time', Ruskin wrote, that buildings themselves came into possession 'of language and of life' (VIII, 234). 'Better the rudest work that tells a story or records a fact, than the richest without meaning' (VIII, 230). It is a crucial acknowledgement of the '"narrativity" of matter' – of how matter is itself 'storied', as Serenella Iovino and Serpil Oppermann have argued (2014, 8); of how 'there is an implicit textuality in the becoming of material formations' (2014, 6), and how, over time, material forms become something more (or even, other) than a function of human ingenuity, human handiwork, or the human urge to make meanings. A built structure becomes, in Ruskin's own words, a 'witness against men'. Herein lay its value; yet here, too, Ruskin argued, society seemed intent on improving, restoring, and erasing that rich, even sacred testimony to human impermanence and inconsequence (VIII, 244).

Wherein lay the source of this scourge? '[B]eautiful things are useful to men', Ruskin later wrote, 'because they are beautiful, and for the sake of their beauty only; and not to sell, or pawn – or, in any other way, turn into money' (1883 Preface to Volume II of *Modern Painters*; IV, 4). But the society of his own day, declared Ruskin, pursued profit, not beauty. The nation was possessed, he insisted, by a 'monetary asceticism, consisting in the refusal of pleasure and knowledge for the sake of money; seen in the present days of London and Manchester' (VII, 424). The pursuit of monetary gain was transforming cities into factories; perhaps the nation as a whole would go the same way, made into

> the furnace of the world; so that the smoke of the island, rising out of the sea, should be seen from a hundred leagues away, as if it were a field of fierce volcanoes; and every kind of sordid, foul, or venomous work which, in other countries, men dreaded or disdained, it should become England's duty to do.
>
> (VII, 425)

As Ruskin well knew, however, the reality was a divided nation, in which 'every kind of sordid, foul, or venomous work' was heaped upon the many for the benefit of the few. As he added, in 'A Joy For Ever' (1857), those few then turned their riches into buildings that celebrated their worship of

'Mammon or Moloch ... the merchant rising to his Mammon matins with the self-denial of an anchorite' (XVI, 138).

By the time, therefore, that Ruskin came to publish the fifth and final volume of *Modern Painters* in 1860, his analysis of art and architecture had already taken on a political dimension, rooted in a strong sense that the true health and wealth of a nation – its 'power and happiness' – were being eaten away by its covetousness, its greed, and its mindless pursuit of 'mechanical productiveness' (VII, 425); by the 'degradation of the operative into a machine', as he had earlier written in 'The Nature of Gothic' (1853), and the environmental inequality which was its concomitant. There is now, he wrote, 'a precipice between upper and lower grounds in the field of humanity, and there is pestilential air at the bottom of it' (X, 194).[6]

Such were, as Ruskin saw them, 'the supposedly progressive tendencies of the modern urban-industrial world' (Macdonald 2012, 134). These threads came together in the essays Ruskin was commissioned to write for the Cornhill Magazine in 1860. Less a break with his previous work than an extension of it, the series was never completed; however, the four essays that he did write, later collected as *Unto this Last*, constitute a powerful statement of Ruskin's views, and a savage excoriation of a profoundly unjust society. As Clive Wilmer puts it, *Unto this Last* is 'first and foremost a cry of anger against injustice and inhumanity' (1985, 21).

For Ruskin, the proximate source of these injustices was the science of political economy, which dominated contemporary thinking about how best to manage a growing industrial economy.[7] In a systematic and carefully researched attack on this supposed science (Batchelor 2000, 175), Ruskin advanced three, overlapping criticisms. His first point was that political economy focused exclusively on the question of how best to maximise wealth, regardless of how that wealth was accumulated: 'moral considerations have nothing to do with political economy' (XVII, 81). Yet it was *im*moral, Ruskin argued, to seek wealth without taking account of the source of it; for making sure that 'this dealing of yours is a just and faithful one', and ensuring that you 'have done your own part in bringing about ultimately in the world a state of things which will not issue in pillage or in death' (XVII, 54). Second, the '*soi-disant* science of political economy' postulated a *homo economicus*, 'economic man', 'a covetous machine', who, abstracted from all forms of 'social affection' (XVII, 25), was motivated solely by self-interest. Thus, society was reduced to an aggregate of individuals driven solely by private gain, where competition took the place of cooperation within an economy anomalously abstracted from that 'human matrix' (Coit 2010, 217). By definition, political economy emerged as an individualistic rather than as a social science (Cockram 2007, 42), neglecting the origins of 'politics' in the Greek *polis*, meaning state, community, citizens (hence the Greek *politikos*) (XVII, 42).[8] Finally, Ruskin argued, political economy failed on its own terms, lacking the rigour, objectivity, and empirical grounding of a science. For example, it failed to define or fully understand what it meant by terms such as wealth, and entirely excluded any consideration of concepts such as intrinsic value or inherent worth (XVII, 135).[9] What Ruskin proposed, in place of this vision of a self-interested society of atomised individuals, was a complex and to some extent contradictory alternate, grounded in a paternalistic, hierarchical vision of society (Coit 2010, 217–18). Not for Ruskin democracy or equality, as Gill Cockram has emphasised (2007, 43). Nonetheless, Ruskin did challenge all those aspects of a *laissez-faire* political economy that led to injustice and hence *in*equality. Moreover, he conceptualised wealth in a way that reflected his growing sense that human and nonhuman life-worlds were mutually interdependent and not therefore separable, which in turn opened up the possibility of a kind of radical democracy in which nonhuman and human were given equal consideration (if not, by any means, equal political participation). Above all, Ruskin strove to lay the ground for a moral economy: 'in order to grow rich scientifically', he wrote, 'we must grow rich justly; and, therefore, know what is just; so that our economy will no longer depend merely on prudence, but on jurisprudence' (XVII, 62).[10]

How, then, does Ruskin make his case? Its defenders claimed that political economy was 'simply the science of getting rich' (XVII, 43), Ruskin argued, and a science it was: if its rules were followed, riches followed, as '[e]very capitalist of Europe' knew (43). In a society of political economy's making, freed from all forms of regulation, intervention, or imposition, everyone was free to be and could be rich – in theory. (The modern version of this fiction is that free-market fundamentalism will make everyone wealthy, or at least, make everyone wealthy eventually, when the wealth has finally trickled down to the poor.) To the contrary, Ruskin argued, wealth acted 'only through inequalities or negations of itself' (XVII, 44). '[T]he art of making yourself rich', wrote Ruskin, was 'equally and necessarily the art of keeping your neighbour poor' (XVII, 44). 'Political economy' was much better defined, he claimed, as *mercantile* economy, since the etymology of the word 'mercantile' (from *merces*, meaning 'pay' [XVII, 44–5]) 'signifies the accumulation, in the hands of individuals, of legal or moral claim upon, or power over, the labour of others' (45). To have power requires poverty (XVII, 45–6); thus, the 'the art of becoming "rich" ... is "the art of establishing the maximum inequality in our own favour"' (XVII, 46).

By contrast, a true political economy, Ruskin argued, was an economy 'of citizens', consisting 'simply in the production, preservation, and distribution, at fittest time and place, of useful or pleasurable things ... adding continually to the riches and well-being of the nation' (XVII, 44). This led him to a discussion of what, in fact, constituted real wealth, which required a definition of what was valuable, a word whose roots Ruskin traced to '*valere*, to be well or strong':

> To be 'valuable', therefore, is to 'avail towards life'. A truly valuable or availing thing is that which leads to life with its whole strength. In proportion as it does not lead to life, or as its strength is broken, it is

> less valuable; in proportion as it leads away from life, it is unvaluable or malignant.
>
> (XVII, 84)

This was a crucial equation, as Dinah Birch has emphasised, and one from which Ruskin derived 'the central rallying cry' of the book (Birch 2012, 32): that 'THERE IS NO WEALTH BUT LIFE' (capitals in the original; XVII, 105); wealth, as opposed to '"illth", causing various devastation and trouble around them in all directions' (XVII, 89); wealth, as the possession of the useful, 'accumulative of capacity as well as of material' (XVII, 88). 'The real science of political economy', insisted Ruskin, 'is that which teaches nations to desire and labour for the things that lead to life: and which teaches them to scorn and destroy the things that lead to destruction' (XVII, 85). Wisdom lay in the pursuit of 'eternal fulness', not death, waste, and 'eternal emptiness' (XVII, 85); it lay in the pursuit of all that was life-enhancing (Cockram 2007, 54).

Tilting first at Jeremy Bentham, then at Thomas Malthus, David Ricardo, and finally John Stuart Mill, Ruskin related his thinking to the fundamental problem with which *Unto this Last* was concerned: how to better a society whose prosperity might be imagined but whose poverty (and impoverished environment) was all too real. What society should indeed seek, Ruskin argued with an ironising eye on Bentham's formulations, was 'the greatest number of noble and happy human beings' (XVII, 105); but how was that possible, when Malthus had raised the spectre of over-population, which only 'hunger, plague, or war' could keep in check (105)? Was this not also the reason why society should pay its workers, not a just wage, but a subsistence one (XVII, 106), as Ricardo had argued? To raise wages would simply encourage the poor to have larger families, and so impoverish themselves all over again; or perhaps it would simply encourage them to drink more (XVII, 106). But why should this be so, Ruskin inquired:

> Suppose it were your own son of whom you spoke, declaring to me that you dared not take him into your firm, nor even give him his just labourer's wages, because if you did he would die of drunkenness, and leave half a score of children to the parish. 'Who gave your son these dispositions?' – I should enquire. Has he them by inheritance or by education? By one or other they *must* come; and as in him, so also in the poor.
>
> (XVII, 106)

As Ruskin asked, why should a whole section of society be improvident, irresponsible, as the political economists assumed? 'Either these poor are of a race essentially different from ours', Ruskin replied, 'and unredeemable (which, however often implied, I have heard none yet openly say), or else by such care as we have ourselves received, we may make them continent and sober as ourselves' (XVII, 106).

Later in the century, a bastardised form of Darwinian theory would indeed lead to claims that the poor were a separate, degenerate species, but as I discussed in Chapter 3, a deterministic narrative was already at work in shaping society's response to the working classes (Hamlin 1998, 187). As Ruskin recognised, it was a widely held belief that poverty was not only the fault of the poor – and 'to be short of money was somehow immoral' (Batchelor 2000, 172) – but a kind of infection that might spread to society as a whole (Hilton 2006, 581). Ruskin's own point was very clear: the poor were no different to the rich, or perhaps their betters, 'holier than we who have left them thus' (XVII, 106).

There remained, as Ruskin accepted, the troubling question of a rapidly growing population, and beyond that, what we would now describe as the carrying capacity of those ecosystem services on which all life depends. 'What end can there be for them [the "multitudes"] at last', asked Ruskin, 'but to consume one another?' (XVII, 108). True, Ruskin added, it would

> be long before the world has been all colonised, and its deserts all brought under cultivation. But the radical question is, not how much habitable land is in the world, but how many human beings *ought* to be maintained on a given space of habitable land.
>
> (italics added; XVII, 108)

Once again, Ruskin invoked the ethical dimension of his argument; once again, Ruskin turned the argument back on his imagined interlocutors. This time, however, the political economists had pre-empted his conclusion, as Ruskin was aware. In the *Principles of Political Economy* (1848), on which Ruskin focused his criticism, John Stuart Mill acknowledged that economic growth must eventually subside into a stationary state (Mill 1852, 316), but insisted that this need not end in a Malthusian disaster: 'prudence and public opinion', Mill argued, might after all 'be relied on for restricting the coming generation within the numbers necessary for replacing the present' (1852, 318). Indeed, Mill unexpectedly conceded, a stationary state might be better than one in which, as at present, 'the normal state of human beings is that of struggling to get on' (1852, 318).

> Most fitting, indeed, is it, that while riches are power, and to grow as rich as possible the universal object of ambition, the path to its attainment should be open to all, without favour or partiality. But the best state for human nature is that in which, while no one is poor, no one desires to be richer …
>
> (Mill 1852, 319)

It is a startling acknowledgement, and it led Mill to the same question that Ruskin posed ('not how much habitable land is in the world, but how many human beings ought to be maintained on a given space of habitable land'

[XVII, 108]). 'A population may be too crowded, though all be amply supplied with food and raiment', acknowledged Mill; '[a] world from which solitude is extirpated, is a very poor ideal' (1852, 321). 'Nor is there much satisfaction in contemplating the world with nothing left to the spontaneous activity of nature', Mill continued:

> with every rood of land brought into cultivation which is capable of growing food for human beings; every flowery waste or natural pasture ploughed up, all quadrupeds or birds which are not domesticated for man's use exterminated as his rivals for food, every hedgerow or superfluous tree rooted out, and scarcely a place left where a wild shrub or flower could grow without being eradicated as a weed in the name of improved agriculture.
>
> (1852, 321)

As Mill concluded:

> If the earth must lose that great portion of its pleasantness which it owes to things that the unlimited increase of wealth and population would extirpate from it, for the mere purpose of enabling it to support a larger, but not a better or a happier population, I sincerely hope, for the sake of posterity, that they will be content to be stationary, long before necessity compels them to it.
>
> (Mill 1852, 321)

Mill's observations anticipate late-twentieth-century debates about the pressures of population growth and the question of long-term environmental sustainability; as John Parham points out, it is on the basis of Mill's chapter 'On the Stationary State' – from which this passage is drawn – that the *Principles of Political Economy* has itself 'long been regarded as a key text in the Western historical development of ecology' (2018, 14). This is also, as we have seen, the same issue with which Ruskin is concerned: was society to continue its unlimited growth until, as Mill put it, there was nothing left 'to the spontaneous activity of nature' (1852, 321)? Ruskin's own response seems, if anything, complacent. He professed himself unconcerned. The world as a whole could never become a factory or a mine, even if England chose that fate for itself (XVII, 110). Some part of the planet would always be given over to cultivation, for that world to be fed; and cultivation required 'a maximum of pure air, and of pure water. Therefore: a maximum of wood, to transmute the air, and of sloping ground, protected by herbage from the extreme heat of the sun, to feed the streams' (XVII, 110). Thus, the earth's cultivation would necessarily create a flourishing biosphere. In other words, Ruskin assumed an essentially benign relationship between the land, and the act of farming it; he assumed that, as David Fairer has argued, 'the underlying georgic premise that we are living in nature's context' necessitates an ecological mindfulness, and with it, a recognition that the land's resources are not inexhaustible (2011, 209). But what of the impact, Ruskin also asked, of 'a mechanical agriculture' (XVII, 110), an agriculture of the kind that Victorian forms of 'high farming' was then pioneering?[11] High farming was an open system, requiring external inputs (for example, of fuel and fertilisers) (Williamson 2002, 139), and as we would now recognise, it anticipated the intensive, agrilogistical forms of farming that have since transformed much of England's farmland into ecologically impoverished monocultures. Perhaps Ruskin was already aware that this was a possibility; nevertheless, he pinned his hopes on the wisdom he sought to foster; for 'a wise population implies the search for felicity as well as for food' (XVII, 111).

'This is', noted E. T. Cook and Alexander Wedderburn in their introduction to *Unto this Last*, 'the pith of the book' (XVII, lxxxvii). 'Felicity' means joy or bliss, but also well-being; it implies a sense of suitability and appropriateness; and it also contains the sense, now more or less forgotten, of prosperity and of prosperous circumstances (Little, Fowler, and Coulson 1983 [1988], 737). How ought people to live, in relation to the land, but also in relation to the wider, nonhuman world, which human ingenuity had not yet contrived to bring under cultivation? With its powers of 'frost and fire', '[t]he desert has its appointed place and work', Ruskin argued, 'but the zones and lands between, habitable, will be loveliest in habitation'; 'nor can any population reach its maximum but through that wisdom which "rejoices" in the habitable parts of the earth' (XVII, 111). Here, once again, Ruskin stressed his belief in the intimate relationship between land and labour, and the ethic of cooperation and care that he believed would follow from it. Almost immediately, however, he went further, to suggest that even those aspects of the nonhuman, natural world that have no direct bearing on farming are nevertheless valuable to a full and flourishing existence – perhaps even essential to it (Shionoya 2014, 25).

> No scene is continually and untiringly loved, but one rich by joyful human labour; smooth in field; fair in garden; full in orchard; trim, sweet, and frequent in homestead; ringing with voices of vivid existence. No air is sweet that is silent; it is only sweet when full of low currents of under sound – triplets of birds, and murmur and chirp of insects, and deep-toned words of men, and wayward trebles of childhood.

Ruskin's reference to the soundscape of the natural world – and in particular, his reference to birdsong – may well now strike the reader as a striking anticipation of Rachel Carson's great work of modern environmentalism, *Silent Spring* (1962), with its opening evocation of a country scene from which birdsong has been driven out by industrial toxins (Garrard 2012, 1–2). Carson's pastoral overture is, of course, only a part of a much larger discussion of the impact of modern pesticides, but Ruskin's own idyll ('fair in garden' and 'full in orchard') is itself much more than a rhetorical, Romanticised flourish. Rather, it is a step towards a defining statement of what constitutes a flourishing life. 'As the art of life is learned', Ruskin continued,

> it will be found at last that all lovely things are also necessary; – the wild flower by the wayside, as well as the tended corn; and the wild birds and creatures of the forest, as well as the tended cattle; because man doth not live by bread only, but also by the desert manna; by every wondrous word and unknowable work of God.
>
> (XVII, 111)

The alert reader will note the 'ironic similarity' (Macdonald 2012, 128) between this passage, and Mill's discussion of the stationary state, whilst also recognising their profound differences: Ruskin's case for 'the art of life' is the culmination of an argument he has been building over the course of all four essays, and indeed, throughout much of the work of the previous decade (Shionoya 2014); Mill's comments appear as an afterthought in an eloquent defence and elaboration of the very system that Ruskin is intent on dismantling (Wilmer 1985, 21).[12] In its place, as we have seen, Ruskin proposed a new science of political economy, whose purpose was to teach 'nations to desire and labour for the things that lead to life' (XVII, 85), and in so doing advance the 'deeper felicity' (XVII, 112) of the human race. This was the source of true prosperity, a prosperity derived from and dependent on the home: for 'all true economy is "Law of the house"' (XVII, 113).

It is an important reference. A few years later (1866), Ernst Haeckel would derive from Darwin's theory of evolution a new form of study that encompassed the complex entanglements of human and nonhuman, biotic and abiotic. He called it 'ecology' (or 'Oecologie') (Haeckel 1866, 286), drawing on the same root to which Ruskin related his point: *oikos*, the house or home, whose constitutive relationships form the basis of all life. Haeckel's object was to define a new kind of science; Ruskin's, to define a new science of life, alert to the wider consequences of every intra-action in a domestic sphere that nevertheless extended beyond the self and species. Ruskin's point is both ecocentric and sociocentric: it speaks to the injustice that distorted the lives of the less fortunate in Victorian society, confined to slum or factory quarter, but also to the environment beyond their own, immediate environs. Ruskin insisted (as he restated, in conclusion) that 'the rule and root of all economy [is] that what one person has, another cannot have', but he also argued that, for a true economy to prevail, whatever was 'used or consumed' must be thought of not only in terms of 'so much human life spent' (XVII, 113); it must also be thought of in terms of its wider consequences; of whether each use or consumption would result 'in the saving [of] present life, or gaining more, [or in] life prevented, or so much slain' (113). The art of life, Ruskin insisted, lay in recognising and responding to what was close at hand – the iniquities on which society was founded – and what lay at the limits of visibility and even of comprehension: '[a]s the art of life is learned', to return to his own words (above, XVII, 111), so every wild flower, wild bird, and desert would be found necessary, as each in themselves valuable (and, at last, valued) elements in that wider network of co-dependencies on which all life depends. Again, and as Ruskin stressed in these essays, 'a *divina scientia*' (XVII, 92) is one that 'teaches nations to desire and labour for the things that lead to life: and which teaches them to scorn and destroy the things that lead to destruction' (XVII, 85); to dismiss any political economist foolish enough to declare 'precious and beneficent things, such as air, light, and cleanliness, to be valueless', or 'the conditions of their own existence, by which alone they can truly possess or use anything ... to be prudently exchangeable, when the markets offer, for gold, iron' (XVII, 85); to recognise, above all, how 'the service of Death, the Lord of Waste, and of eternal emptiness, differs from the service of Wisdom, the Lady of Saving ["Madonna della Salute", "Lady of Health"], and of eternal fulness' (XVII, 85). Recognising (and acting on) these distinctions, Ruskin insisted, would lead to the 'true felicity of the human race' (XVII, 111), a felicity, prosperity, and happiness alert to 'every wondrous word and unknowable work of God' (XVII, 111).

In Ruskin's opinion, it is the very unknowability of those works that requires their prudent and careful consideration, and a way of life (suited *to* life) 'resolved to seek – not greater wealth, but simpler pleasure; not higher fortune, but deeper felicity; making the first of possessions, self-possession' (XVII, 112). His philosophy implies a care and caution entirely compatible with today's late modern articulations of sustainability. It is, however, still more radical in its implications for the economic basis of human existence. In the midst of an emerging consumer culture, Ruskin asked whether consumption was in or of itself good, or whether, as he concluded, any kind of 'luxury' – any kind of wealth, as capitalism constructs society's sense of it – could ever be desirable or even guiltless, 'if we saw clearly at our sides the suffering which accompanies it in the world' (XVII, 114).[13]

*

As this discussion illustrates, *Unto this Last* takes as its starting and end point the suffering of the poor, condemned to impoverished lives under the conditions created by industrial modernity and underpinned by a free-market economy. As Ruskin wrote as early as 1853,

> the great cry that rises from all our manufacturing cities, louder than their furnace blast, is all in very deed for this, – that we manufacture everything there except men; we blanch cotton, and strengthen steel, and refine sugar, and shape pottery; but to brighten, to strengthen, to refine, or to form a single living spirit, never enters into our estimate of advantages.
>
> (X, 196)

In *Unto this Last*, he therefore set out to disabuse his contemporaries of their complacent belief that a system or 'science' of political economy offered the best of all possible responses to the advent of industrial modernity. Instead, he sought to put in its place 'a system based on mutual interdependence', in which collective welfare took priority over individual aggrandisement (Cockram 2007, 43). In making his argument, Ruskin asked profound questions about what constituted a flourishing existence, and how best it might be achieved; as he recognised, each individual life was inextricably caught up in 'the infinitely varied worlds of the Other' (Levine 2000, 82). For humans to flourish, so must the nonhuman world. Ultimately, society and nature were inseparable, as linked expressions of God's will. Thus, *Unto this Last* represents both a radical response to the Condition-of-England Question first posed by Carlyle, and a tentative expression of an emergent justice, in which ecocentric concerns are integral to the construction of an environmentally just society.

It was, however, the sociocentric, perhaps even socialist emphasis of the book that caught the attention of Ruskin's contemporaries (Levine 2000, 76), and 'raised a storm of indignant protest' (Cook 1912, 6). His publishers declared that the papers were 'deeply tainted with socialistic heresy', claimed E. T. Cook (1912, 8); the series was cut short.[14] For Ruskin, it was a personal turning point. Having won widespread acclaim and respect for his art criticism, he now found himself 'rendered helpless' by the same middle-class readers whose good opinion he had once courted (Cockram 2007, 45). When, two years later, he published the Cornhill papers as *Unto this Last*, the volume sold poorly. A thousand copies were printed; 11 years later, over 100 remained unsold (Cook 1912, 13). Why was this so? Cook suggested that the principles of political economy were then so well-established – so firmly entrenched – that they had come to seem natural (1912, 11), just as today, the neoliberal insistence on the primacy of the free market has itself come to seem like an unassailable orthodoxy, so commonplace that it is rarely even identified *as* neoliberalism (Monbiot 2016 [2017], 3). Yet dominant as the principles of political economy may have seemed, they were nonetheless vulnerable to precisely the kind of moral criticism that Ruskin made. Was it right that society was so unbalanced? Was it right that a substantial section of the population lived in conditions of such squalor, whether or not they were in work? What good did wealth do for a nation that could not even create the right conditions for its people to live full and meaningful lives? All too obviously, these principles benefited the few at the expense of the many, as the fortunate few themselves recognised. When *The Saturday Review* declared Ruskin's efforts to equalise the relations of rich and poor 'repulsive' (qtd in Cook 1912, 7) – when it declared that it would not be 'preached to death by a mad governess' (qtd, 6) – it did so not from a position of strength, but from a growing recognition of its own weakness: if Ruskin's incendiary opinions were not crushed, wrote the *Manchester Examiner and Times*, 'a moral floodgate may fly open and drown us all' (qtd in Cook 1912, 13).

It would not do so in Ruskin's own lifetime. Increasingly marginalised by a 'resistant audience' (O'Gorman 2012, 21), Ruskin eventually lapsed into silence. Yet *Unto this Last* steadily gained in popularity as the century wore on (Cockram 2007, 90). Indeed, it was cited as a significant influence on the first-generation of Labour parliamentarians (Batchelor 2000, ix; Eagles 2007, 95). What they took from it is another matter.[15] Then as now, *Unto this Last* eludes conventional political interpretations, and the alignments of left and right. As Frost underlines, it is a work of quixotic brilliance that can only be appreciated in the context of the 'interpenetration of organic, cultural, and political discourses' which characterises Ruskin's mature work (Frost 2014, 38); it must be seen in terms of both its ecocentric and its sociocentric charge. Ruskin's influence was profound, but like many of the writers whose work I have explored in this book, the implications of his thinking have not yet been fully worked out. What, then, should we make of Ruskin and his contemporaries, who responded so forcefully and imaginatively to the environmental inequalities that the Victorians were bequeathing to the world? These are the questions to which I turn in the final, concluding chapter, which traces the impact of these thinkers into the later Victorian period and beyond, and reconsiders their ideas in light of our own emerging views about what should constitute environmental justice in the twenty-first century.

*

Notes

1 This notwithstanding, the definitive edition of Ruskin's work – the Library Edition edited by E. T. Cook and Alexander Wedderburn – is digitally available from the website of Lancaster University (www.lancaster.ac.uk/the-ruskin/the-complete-works-of-ruskin/). At the time of its publication (1903–12), however, this 39-volume opus seems to have 'acted like a tombstone on Ruskin's reputation' (Clark 1963 [1983], xii).

2 As I highlighted in the Introduction, first-wave ecocriticism tended to focus on nature writing, Transcendentalists, and Romantics, and on figures such as Emerson, Thoreau, and Wordsworth; what many regard as a foundational work of British ecocriticism, Jonathan Bates's *Romantic Ecology* (1991), is subtitled 'Wordsworth and the environmental tradition'. But as John Parham points out (2002, 156), Bate devoted a substantial part of the book (Bate 1991, 58–64) to a discussion of Ruskin's proto-ecological ideas. Four years later, Michael Wheeler published his edited collection *Ruskin and environment: the storm-cloud of the nineteenth century* (1995). Ruskin has, it may be said with some fairness, been recognised from the first days of ecocriticism as a notable green thinker (Wheeler 1995, 2–3; MacDuffie 2014, 137), and perhaps even, as Laurence W. Mazzeno and Ronald D. Morrison suggest, 'the most "ecological" writer among the Victorians' (2017 [2019], 6).

3 Cook and Wedderburn's Library edition is the copy text for this chapter, and hereafter, all references to this work will take the simplified format of volume and page number.

4 Ruskin was referring to selective breeding; today, genetic engineering and modification has taken this a step further. Arguably, these acts of interference are themselves forms of violence, and as such, 'part of a larger repertoire of strategies that "put nature to work"' (Moore 2016, 5).

5 The related point, relevant to an analysis which takes new materialist thinking as a touchstone, is the extent to which Ruskin was himself a materialist. As Clive Wilmer insists, Ruskin remained 'a Christian, even after his "unconversion"' (1996, 86); Ruskin opposed the work of Darwin, George Levine adds, 'because he found Darwin's view that it was matter all the way down intolerable and immoral' (2008, 223). But as Frost points out, this theory of Ruskin's 'virulent opposition to Darwin and materialism' (2011, 367) tends to overlook the extent to which, when Ruskin's work is viewed as a whole, Ruskin understood and acknowledged 'the dynamic materialities of environment' (Frost 2011, 369); his views are, as Emily Coit points out, 'suffused' with vitalism (2010, 235).

6 Ruskin's reference to 'pestilential air' may be taken literally, as a reference to the miasmatic theory of disease transmission (see Chapter 2), but also as reflection 'of a broader disordering brought about by deliberate human interference with vital helpfulness' (Frost 2017 [2019], 25). The corollary of that emphasis on pollution was Ruskin's recognition that a healthy, natural environment was one that also made healthy and happy those who inhabited it. Frost (2017 [2019], 24–6) highlights the passages in *The Crown of Wild Olive* (1873) where Ruskin describes the 'sources of the Wandel' (386), 'the Croydon stream that was a childhood playground' (Frost 2014, 16), and its later despoliation, turned into a dumping ground for 'heaps of dust and slime, and broken shreds of old metal, and rags of putrid clothes' (XVIII, 386); yet 'God meant those waters to bring *joy and health*' (italics added; XVIII, 386). Ruskin goes on to write that '[h]alf-a-dozen men, with one day's work could cleanse those pools, and trim the flowers about their banks, and make every breath of summer air above them rich with cool balm; and every glittering wave medicinal' (XVIII, 387). 'But that day's work is never given' (XVIII, 387), Ruskin added, instead drawing attention to the railings he has seen put up in a Croydon street, intended, perhaps, to dignify the establishment, but instead contriving to create a piece of dead ground into which passers-by can and do hurl their rubbish – and from which that refuse cannot be retrieved (XVIII, 387). The very making of these railings took work – and involved suffering, Ruskin stresses – yet the result was 'an entirely (in that place) valueless, piece of metal, which can neither be eaten nor breathed, instead of medicinal fresh air and pure water' (XVIII, 388) (note again the emphasis on the medicinal properties of the springs and their environs). So why has effort been expended on one (the railings) and not the other (the springs)? 'There is but one reason for it, and at present a conclusive one, – that the capitalist can charge percentage on the work in the one case, and cannot in the other' (XVIII, 388). The related point, in the language of modern economics, is that capitalism treats as externalities anything to which a price-tag cannot be readily attached, 'medicinal fresh air and pure water' being two of them (XVIII, 388). Ruskin would later instigate a project to clean up the spring (Frost 2014, 73–80).

7 As Clive Wilmer points out (1985, 19–21), Ruskin's targets included T. R. Malthus, who argued that a growing population would be curbed only by starvation, and the pioneering Utilitarian thinker, Jeremy Bentham, whose attempts to ground an ethical framework in 'the greatest happiness of the greatest number' drew widespread criticism. Their thinking also informed the empiricist tradition of classical economics founded by Adam Smith, continued by David Ricardo and James Mill, and elaborated by Mill's son, John Stuart Mill; it is their science of 'Political Economy' on which Ruskin focuses.

8 The state, the political economists argued, has no role in creating wealth – and none in monitoring, regulating, or controlling its creation. Its role was rather to absent itself; to let the market alone (hence '*laissez faire*'). A free market would, in turn, make for a free society, fusing the economic with the political in a progressivist utopia, an argument that would resurface under the guise of 'neoliberalism' in the 1970s and 1980s, prompting a decisive shift away from more collectivist, communal forms of policy-making that sought to level out inequality through state intervention and regulation of the market. Today, policy making in many western societies is still dominated by neoliberal arguments, or what its (few) critics call 'free-market fundamentalism' (Monbiot 2016 [2017], 2–3); alternatives from socialism to anarchism or communism are widely demonised (2016 [2017], 3).

9 In his formative work on *The Wealth of Nations* (1776), Adam Smith distinguished between different kinds of value (Smith 1937, 28), but pursued his inquiry only in relation to value in exchange, on which John Stuart Mill would later rest his own definition (Coit 2010, 233); the related question, which lies beyond my scope here, is whether Smith's identification of another kind of value – which he called value in use – is in itself sufficient to capture Ruskin's sense of value as intrinsic, and therefore independent of utility.

10 In the 'Preface' that Ruskin added to the essays when republished in book form in 1862, he added a list of the practical proposals that he derived from them (see XVII, 21–3; they are helpfully glossed by Cook and Wedderburn, XVII, lxxx-vii). As Gill Cockram notes, these proposals included '[f]ixed wages, government subsidised industries, state care for the poor, elderly, and inform, quality control of goods, and state education', proposals which were then sufficient to earn him 'both derision and allegations of socialism' (2007, 43). To Kenneth Clark, writing a century later, these proposals seemed like the unremarkable basis of a social-democratic welfare state of the kind that post-war Britain had itself pioneered (1963 [1983], xviii–xix); to Milton Friedman, writing in the United States at more or less the exact same moment, they were examples of the kind of state interference which was everywhere stifling economic growth and conspiring against the creation of a truly free society (*Capitalism and Freedom* (1962, 6). Clark (1903–83) did not anticipate that, within his own lifetime, the tide of political opinion would swing against the Ruskinian vision of a fair and just society, a point to which I return in the Conclusion.

11 Ruskin's 'mechanical agriculture' and Mills 'improved agriculture' (Mill 1852, 321) – refer to what is today defined as 'high farming', '"high" in every sense of its operations, requiring high capital investment in buildings, machinery, roads, drainage, highly bred livestock, animal feed and new crop strains as well as fertilisers, all usually on larger farm units' (Kinsley 2018, 132); it was, Lesley Kinsley adds, a 'high input, throughput, output, intensive and open system' (2018, 132–3).

12 As Parham observes, '[t]here is little of nature *per se* in "The Stationary State" chapter until Mill introduces a poetic flourish near the end, where, notionally, his "stationary" economics gradually reaches beyond the human alone … The value Mill attaches to unspoiled nature remains largely couched in terms of its benefit to humans' (2018, 17). To this extent, Ruskin's dismissal of Mill is fair. As Parham adds, however, Mill would later develop a more comprehensive and ecologically mindful vision of the interrelationship between human and nonhuman life (2018, 20); 'The Stationary State' chapter is better regarded as the beginning of Mill's thinking on the subject, rather than its end (18).

13 Thus, and as Emily Coit discusses in detail, Ruskin also emerges as 'an important early thinker' in a developing awareness of 'ethical behaviour in the context of consumer culture' (2010, 218).

14 Cook exaggerates, although memorably. Nevertheless, the Cornhill, in which the essays were first published, was squarely aimed at a middle-class readership, and as Tim Hilton argues, 'a first principle' was that articles should be 'uncontroversial' (2000, 12).

15 '[W]e are all socialists now', the liberal politician Sir William Harcourt declared; '[h]e might have said', retorted Cook, that '[w]e are all Ruskinians' (qtd in Cockram 2007, 91). Cook was not implying that we equate socialism with Ruskin's views, however; Ruskin's views were never readily reducible to any single political position. If anything, Macdonald argues, Ruskin's 'advocacy is that of the radical Tory rather than the egalitarian radical' (2012, 135). Cook's point is rather that Ruskin's influence was growing, as his ideas were absorbed into the political mainstream – at least in part. It may only be now, from our late-modern, environmentally-aware perspective that the full implications of his arguments can be realised. For a detailed discussion of Ruskin's influence, see O'Gorman (2012).

Works cited

Alaimo, Stacey. 2010. *Bodily Natures: Science, Environment, and the Material Self*. Bloomington, IN: Indiana University Press.

Barad, Karen. 2007. *Meeting the Universe Halfway: Quantum Physics and the Entanglement of Matter and Meaning*. London: Duke University Press.

Batchelor, John. 2000. *John Ruskin: No Wealth but Life*. London: Chatto & Windus.

Bate, Jonathan. 1991. *Romantic Ecology: Wordsworth and the Environmental Tradition*. Abingdon: Routledge.

Birch, Dinah. 2004 [2009]. 'Introduction'. John Ruskin, *Selected Writings*, edited by Dinah Birch, ix–xxxvi. Oxford: Oxford University Press.

Birch, Dinah. 2012. '"What Is Value?": Victorian Economies of Feeling'. *Carlyle Studies Annual* 28: 31–48. Available at jstor.org, www.jstor.org/stable/26594304. Accessed 19 December 2022.

Clark, Kenneth. 1963 [1983]. *Ruskin Today*. Harmondsworth: Penguin.

Cockram, Gill G. 2007. *Ruskin and Social Reform: Ethics and Economics in the Victorian Age*. London: Tauris Academic Studies.

Coit, Emily. 2010. '"This Immense Expense of Art': George Eliot and John Ruskin on Consumption and the Limits of Sympathy'. *Nineteenth-Century Literature* 65 (2): 214–45. Available at jstor.org, https://doi.org/10.1525/ncl.2010.65.2.214. Accessed 21 December 2022.

Cook, E. T. 1912. *The Life of John Ruskin, Vol. II*: 1860–1900. 2nd edn. London: George Allen & Company.

Cook, E. T., and Alexander Wedderburn (eds). 1903–12. *The Library Edition of John Ruskin's Works*. 39 vols. London: George Allen. Available at www.lancaster.ac.uk/the-ruskin/the-complete-works-of-ruskin/. Accessed 5 January 2023.

Eagles, Stuart. 2007. 'The "insinuating touch of influence": aspects of Ruskin's political legacy'. *Ruskin in Perspective: Contemporary Essays*, edited by Carmen Casaliggi and Paul March-Russell, 83–100. Newcastle: Cambridge Scholars Publishing.

Fairer, David. 2011. '"Where Fuming Trees Refresh the Thirsty Air': The World of Eco-Georgic'. *Studies in Eighteenth-Century Culture* 40: 201–18. doi: 10.1353/sec.2011.0006

Frost, Mark. 2011. '"The Circles of Vitality": Ruskin, Science, and Dynamic Materiality'. *Victorian Literature and Culture* 39 (2): 367–83. Available at jstor .org, www.jstor.org/stable/41307872. Accessed 23 June 2022.

Frost, Mark. 2014. *The Lost Companions and John Ruskin's Guild of St George: A Revisionary History*. London: Anthem Press.

Frost, Mark. 2017 [2019]. 'Reading nature: John Ruskin, environment, and the ecological impulse'. *Victorian Writers and the Environment: Ecocritical Perspectives*, edited by Laurence W. Mazzeno and Ronald D. Morrison, 13–28. Abingdon: Routledge.

Garrard, Greg. 2012. *Ecocriticism*. 2nd edn. Abingdon: Routledge.

Haeckel, Ernst. 1866. *Generelle Morphologie der Organismen, Volume 2*. Berlin: Georg Reimer.

Hamlin, Christopher. 1998. *Public Health and Social Justice in the Age of Chadwick: Britain, 1800–1854*. Cambridge: Cambridge University Press.

Hilton, Boyd. 2006. *A Mad, Bad, and Dangerous People? England 1783–1846*. Oxford: Clarendon Press.

Hilton, Tim. 2000. *John Ruskin: The Later Years*. London: Yale University Press.

Iovino, Serenella, and Serpil Oppermann. 2014. 'Introduction: Stories Come to Matter'. *Material Ecocriticism*, edited by Serenella Iovino and Serpil Oppermann, 1–17. Bloomington, IN: Indiana University Press.

Kinsley, Lesley. 2018. 'Guano, science and Victorian High Farming: An Agro-Ecological Perspective'. *Victorian Sustainability in Literature and Culture*, edited by Wendy Parkins, 126–45. London: Routledge.

Levine, Caroline. 2000. 'Visual Labor: Ruskin's Radical Realism'. *Victorian Literature and Culture* 28 (1): 73–86. Available at jstor.org, www.jstor.org/stable/25058492. Accessed 20 December 2022.

Levine, George. 2008. 'Ruskin, Darwin, and the Matter of Matter'. *Nineteenth Century Prose* 35 (1): 223–49. https://doi.org/10.1017/S1060150317000171

Little, William, H. W. Fowler, and Jessie Coulson. 1983 [1988]. *The Shorter Oxford English Dictionary, Volume I*. 3rd ed. London: Guild.

Macdonald, Graham A. 2012. 'The Politics of the Golden River: Ruskin on Environment and the Stationary State'. *Environment and History* 18 (1): 125–50. Available at jstor.org, www.jstor.org/stable/23250895. Accessed 19 December 2022.

MacDuffie, Allen. 2014. *Victorian Literature, Energy, and the Ecological Imagination*. Cambridge: Cambridge University Press.

Mazzeno, Laurence W., and Ronald D. Morrison. 2017 [2019]. 'Introduction: Practical Ecocriticism and the Victorian Text'. *Victorian Writers and the Environment: Ecocritical Perspectives*, edited by Laurence W. Mazzeno and Ronald D. Morrison, 1–12. Abingdon: Routledge.

Mill, John Stuart. 1852. *Principles of Political Economy: with some of their applications to social philosophy, vol. II*. 3rd edn. London: John W. Parker and Son. Available at archive.org, https://archive.org/details/in.ernet.dli.2015.180870/page/n9/mode/2up?view=theater. Accessed 27 May 2022. First published in 1848.

Monbiot, George. 2016 [2017]. *How Did We Get Into This Mess? Politics, Equality, Nature*. London: Verso.

Moore, Jason W. 2016. 'Introduction'. *Anthropocene of Capitalocene? Nature, History, and the Crisis of Capitalism*, edited by Jason W. Moore, 1–11. Oakland, CA: PM Press.

O'Gorman, Francis. 2012. '"Influence" in the Contemporary Study of the Humanities: The Problem of Ruskin'. *Carlyle Studies Annual* 28: 5–30. Available at jstor.org, www.jstor.org/stable/26594303. Accessed 19 December 2022.

Parham, John. 2002. 'Was there a Victorian Ecology?' *The Environmental Tradition in English Literature*, edited by John Parham, 156–71. Aldershot: Ashgate.

Parham, John. 2018. 'A Not So 'Stationary State': John Stuart Mill's Sustainable Imagination'. *Victorian Sustainability in Literature and Culture*, edited by Wendy Parkins, 14–31. London: Routledge.

Plumwood, Val. 1993 [1997]. *Feminism and the Mastery of Nature*. London: Routledge.

Rose, Phyllis. 2003. 'Ruskin's Power'. *The American Scholar* 72 (2): 87–95. Available at jstor.org, www.jstor.org/stable/41221124. Accessed 19 December 2022.

Shionoya, Yuichi. 2014. 'Ruskin's Romantic Triangle: Neither Wealth nor Beauty buy life'. *History of Economic Ideas* 22 (1): 15–49. Available at jstor.org, www.jstor.org/stable/43924191. Accessed 19 December 2022.

Smith, Adam. 1937. *The Wealth of Nations*, edited by Edwin Cannan. New York, NY: Random House. Available at archive.org, https://archive.org/details/in.ernet.dli.2015.207956/page/n5/mode/2up?view=theater. Accessed 21 June 2022.

Wheeler, Michael. 1995. 'Introduction'. *Ruskin and Environment: The Storm-Cloud of the Nineteenth Century*, edited by Michael Wheeler, 1–9. Manchester: Manchester University Press.

Williams, Daniel. 2016. 'The Clouds and the Poor: Ruskin, Mayhew, and Ecology'. *Nineteenth-Century Contexts* 38 (5): 319–31. doi: https://doi.org/10.1080/08905495.2016.1219197

Williamson, Tom. 2002. *The Transformation of Rural England: Farming and the Landscape, 1700–1870*. Exeter: University of Exeter Press.

Wilmer, Clive. 1985. 'Introduction'. *John Ruskin, Unto This Last and Other Writings*, edited and introduced by Clive Wilmer, 7–37. Harmondsworth: Penguin.

Wilmer, Clive. 1996. 'Was Ruskin a materialist?' *Time and Tide: Ruskin and Science*, edited by Michael Wheeler, 85–97. London: Pilkington Press.

Conclusion
Looking forward

What is it to live in an environmentally just society, in which every citizen has an equal opportunity to live full and flourishing lives? This question began to take on its modern meaning during the Victorian period, when it became obvious that the lives of Britain's working classes were increasingly distorted by the reality of environmental injustice, a reality all too often discounted as the unfortunate but unavoidable side-effect of an age of improvement. What was the answer to the all too palpable impress of an 'abnatural' environment (Taylor 2016, 5), in which the very conditions of existence were beset with often unseen, even unparalleled new risks? For Matthew Arnold (1822–88), one of the most celebrated critics of the later Victorian period, the answer lay in rising above these pressing new realities, and looking within for perfection: it lay in cultivating 'our humanity proper, as distinguished from our animality' (Arnold 1965, 94), and developing a 'best self' (134–5) disentangled from the hazards of the age. True culture, Arnold wrote in *Culture and Anarchy* (1867), lay in a 'pursuit of sweetness and light' (1965, 99) that would transcend 'the mechanical and material' (95). As others were more ready to acknowledge, however, 'material well-being' (Arnold 1965, 105) was itself fundamental to a flourishing existence, and a flourishing existence in turn depended on the nature of the environment that society preserved or cultivated and which it made available to its people. This was not just a question of what people needed – the basic means of subsistence, say – but of subtler demands for the goods that constituted a full life. As John Ruskin had argued in *Unto this Last* (1860) the nonhuman and natural world was in itself of value, and as such deserved its own opportunity to flourish; as he also recognised, a flourishing natural world was necessary to complement, perhaps even complete human existence. 'No air is sweet that is silent', Ruskin insisted; 'it is only sweet when full of low currents of under sound – triplets of birds, and murmur and chirp of insects … all lovely things are also necessary' (Cook and Wedderburn 1903–12, Vol. XVII, 111).

Ruskin was not alone in drawing this conclusion. For the early Victorians whose work I have discussed in this book, the material world mattered, and mattered in ways that industrial modernity had thrown into sharp relief at the very moment it threatened to alter and impoverish that world irrevocably. No theory of evolution or science of ecology (still less the new materialist

DOI: 10.4324/9780367821609-8

thinking of our own late-modern times) was needed to highlight the constitutive co-dependencies that defined human and nonhuman flourishing: a newly industrialised world was one that made plain the importance of these interconnections – these intra-actions – if only because it substituted a denatured or abnatural world for one in which the nonhuman and natural was a living, agential presence. Thus, a developing appreciation of social injustice was inseparable from a wider awareness of environmental injustice, conceptualised in ways that did not assume 'the environment' existed solely (if at all) as an economic externality. Environmental justice was emergent justice, in the strong sense that a fair and just society was one in which (intrinsic) value was distributed along and through these networks of connectivity and co-dependency.

Yet these emerging ideas about how to constitute a flourishing and fair society did not translate into substantive, structural change in the later years of Victoria's reign. It was Thomas Carlyle, not Friedrich Engels or Karl Marx, who correctly identified what would come of the 'Condition of England' and the unrest it generated in the 1840s: not revolution, but accommodation, compromise, and reform.

> Mill-owners, Manufacturers, Commanders of Working Men … admonished by Trades-unions, Chartist conflagrations, above all by their own shrewd sense kept in perpetual communion with the fact of things, will assuredly reform themselves, and a working world will still be possible …
>
> (Carlyle 1843 [1897], Vol 6, 387)

Carlyle was correct: a working world was indeed possible, with work much in evidence, but it was one marked by the growing separation of 'the Privileged and the People' (Disraeli 1981 [2008], 245). What the early Victorians had pioneered, the later Victorians extended and consolidated: increasingly, those with the means to do so migrated to the suburbs, driven there by the desire 'to escape the "contaminating" influence of the urban environment' (Horn 1997 [1999], 21). As Bill Luckin observes, it was a revolution (2015, 172), and it was reproduced all across the country (Steinbach 2017, 13).[1] It was not simply a matter of escaping an unwholesome environment, but of taking advantage of a healthy one. As enterprising speculators recognised, people did indeed take real pleasure in contact with a 'medicinal' nonhuman and natural world (Cook and Wedderburn 1903–12, Vol. XVIII, 388), and their own commercial success depended on providing it. Garden suburbs like the pioneering Bedford Park – constructed from 1875 onwards, and naturally situated towards the less-smoky west of London – were carefully contrived to combine a leafy, landscaped world with easy access to the city, whilst out of reach of its baleful influences (Bédarida 1970 [1985], 102; Bilston 2019, 7). Suitably packaged as just this kind of green space, Bedford Park was advertised as 'the healthiest place in the world (Annual Death Rate under 6 per Thousand)' (1882; Liebreich and Cruz, 2018).

'[M]odern society acknowledges no neighbour', wrote Disraeli in *Sybil* (1981 [2008], 65); polite society, certainly. Safely ensconced in these leafy, hybrid spaces, the fortunate could and did pretend that theirs was the best of all possible worlds, an act of denial that Charles Dickens christened 'Podsnappery' (1864–5 [1997], Ch. 11). '[T]here is not a country in the world', insists Mr. Podsnap in *Our Mutual Friend* (1864–5), 'where so noble a provision is made for the poor' (Dickens 1864–5 [1997], 143), in reference to the thoroughly *ig*noble Poor Law. Dickens's portrayal of the Podsnaps verges on caricature, but it was nonetheless grounded in fact: the Podsnaps embodied what the writer and journalist Richard Jefferies described in 1883 as 'an interested ignorance, kept up by strenuous exertions' (1907 [1913], 176–7). Henry James offered a more elegant version of the same self-serving logic of denial. 'London was hideous, vicious, cruel' (1893, 10), he wrote in his essay on the city (1888), and if 'a genial summary' (27) of this 'murky modern Babylon' (1) was to be made, it was first necessary to ignore the '[t]he uglinesses, the "rookeries," the brutalities' (27).

Thus, and whilst the rich retired to their villas, there to enjoy the amiable fiction of a great and civilised metropolis, 'decent and desperate people' (Hall 2002 [2003], 19) were left to make their way in an increasingly polluted city. By the century's end, the separation of rich and poor was so total that the slums had taken on an other-worldly dimension (Cubitt 2019, 3–4). As social crusaders like Charles Booth pointed out, they now constituted a kind of '"terra incognita" ... on our social map' (1903, 17). Slum fiction sprang up to populate this blank slate with sometimes salacious and melodramatic tales of the pitiable poor; writers such as Arthur Morrison worked, in part, to supply more balanced narratives of slum life's 'quiet desperation and silent suffering' (Cubitt 2019, 3). But slum fiction was not the only new literary development that enabled writers to engage with this divide, or imagine what might come of it. The rapid growth of speculative – or science – fiction allowed writers to consider the wider impacts of industrial modernity, both in the long and short term, and explore what Booth described as the 'ignorant unnecessary fears' (1903, 18) that society projected onto the slums.[2] Were the suburbanites really so safe in the suburbs? It was H. G. Wells (1866–1946) who offered the most dramatic answer, in *The Time Machine's* depiction of the Eloi and Morlocks (1895 [2005]), the Eloi an evolutionary extension of the leisured garden suburbanites, the predatory Morlocks an extension of the labouring poor, raised in the Stygian gloom that industry created. But it was a far less well-known writer, Fred M. White (1859–1935), who contrived the more immediately compelling, and no less chilling response, in a short story that drew on the journalistic exposés of London's dust-mounds and wastelands discussed in the Introduction. In 'The Dust of Death' (1903), one of six 'Doom of London' stories, White describes the fate of a fictional garden suburb called (with a nod to the first of its kind) Devonshire Park. Its name has been calculated to appeal to an affluent and aspirational clientele; it hints at aristocratic privilege and a green and pleasant parkland far removed from the plague-spots and fever-dens that, as polite society had only recently rediscovered, still formed the city's nucleus. But Devonshire Park hides a secret. As I highlighted in the Introduction, the dust-mounds of London were part of a circulatory economy; they had a value, for example as a component in brick and tile manufacture. It was no great stretch, therefore, to imagine that 'dust' might also be used as the footings for new developments, even as 'the foundations of a suburban paradise' like this one (White 1903 [2020], 49). But as we have also seen, 'dust' was something of a euphemism. The mounds were made up of all kinds of refuse and waste: '[c]ontaminated rags, torn paper, road siftings, decayed vegetable matter, diseased food, fish and bones' (White 1903 [2020], 45), and perhaps even human excrement (Brattin 2002, 27), a detail too offensive for the writers of the day to mention explicitly. As White's narrator explains, Devonshire Park has itself been built on the site of 'deserted brickfields' (White 1903 [2020], 44), and levelled with 'a mass of living matter' (45). Matter is agential, dynamic, and unpredictable: what might this foot-deep strata incubate beneath the 'external beauties of Devonshire Park' (White 1903 [2020], 49)?

Here, White tapped into the kinds of anxiety that mark the emergence of a risk society, in which 'unknown and unintended consequences come to be a dominant force in history and society' (Beck 1992 [2012], 22). When in the 1870s James Greenwood visited the dust-mounds and nuisance neighbourhoods of Belle-Isle, he was suitably appalled by the injustice of the 'pestilent colony' (Greenwood 1873, 67) he encountered, but his reaction also contained a strong vein of self-interest. As he told his middle-class readers, the 'poisoned air' (Greenwood 1873, 69) produced by Belle-Isle's toxic trades and lethal factories – and by the 'fever-dens' (68) in which the poor were living – might yet rebound on those whose wealth had so often been made at the expense of the Belle-Islanders:

> It is not nice to talk about such matters; it was very far from nice to investigate them; but, since such vileness exists, has existed doubtless for years, and will continue to exist for all that the parochial authorities can do to make an end of it, it becomes necessary to expose it for common safety, no less than for mercy's sake. *The risk we run in shirking such questions is incalculable.*
>
> (Italics added; Greenwood 1873, 68)

Greenwood feared that the contamination might carry to well-to-do neighbourhoods: in White's story, the rich have built their paradise on the source of it – the dust, which beneath the surface has acted as 'a veritable hotbed of disease' (White 1903 [2020], 45), creating a new and virulent strain of diphtheria. An outbreak is inevitable: White's story traces its trajectory, as

Devonshire Park is overwhelmed by a modern plague of (high) society's own making. At first, science is helpless to respond; under the conditions created by industrial modernity, as Beck predicted (1992 [2012], 29), new risks test the very limits of scientific comprehension. Thousands die before a cure is found. By then, however, a no less awful fate – for so the narrative depicts it – has befallen the suburbs: as the rich have fled to safety, the 'slum people' (White 1903 [2020], 56) have taken up residence. They have no fear of disease; 'they are too familiar with it for that' (White 1903 [2020], 53); and so, adds the narrator with a nod to the rookeries of Dickens' *Bleak House* (1852–3), '[t]he smiling paradise [became] a huge estate in Chancery' (54).

The reader may be surprised to find, however, that this is not the cue for a thundering, Dickensian tirade against society's self-interestedness, and its blindness to the iniquities it has perpetuated. White's narrative is focalised through a society doctor appalled by this 'ragged invasion' (White 1903 [2020], 55), this occupation 'by Whitechapel at its worst' (54). For White's 'distinguished physician' (White 1903 [2020], 41), in other words, the real risk is that society's natural (or naturalised) order has been upset, and the propertied displaced by the dispossessed; 'creeping westward' (53) from their haunts in the East End, the 'outcasts' (54) are themselves the plague. But perhaps the real interest of White's story lies not with its suggestion that the poor are themselves now regarded as a pollutant, still less the suggestion that the rich and privileged have got their just (and justly poisonous) desserts; it rests on the fact that, by virtue of their wealth, most of the rich escape their fate. It is they who can call on the best doctors for help and the leading scientists for a cure, and they whose wealth enables them to leave, *en masse*, as they left for the garden suburbs in the first place, carrying themselves ever further from the threats engendered by industrial modernity. Truly, 'money can command everything' (White 1903 [2020], 52). Thus, White's tale returns the reader to its own beginnings, in a divided society, where pollution and poverty are coincident, and the rich continue to find ways in which to flourish, far from the source of their wealth and the exploitation on which it is built.

The rich will always be with us, White's story suggests: but need that be the case? As I have highlighted, society's critics were not slow to identify the root cause of the social and environmental inequality that beset the nation: it lay in what Matthew Arnold described as 'the present great movement towards wealth and industrialism' (1965, 104). They recognised, in other words, that the source of these ills and iniquities lay in the emergence of industrial modernity, and the accelerating and increasingly globalised process of capital accumulation. To some, conflict seemed inevitable, and necessary. 'Either the weed must strangle the flower, or the flower must destroy the weed', wrote the Chartist leader Ernest Jones in 1852 (1). 'Either the employing class must be ruined by the employed, or the employed by the employer', Jones added; '[s]ocial war' was the only answer (1852, 1). But with the failure of Chartism around the middle of the century, revolutionary sentiments gave way to a reformist belief that a divided society could be fixed from within, through the power of a greatly expanded state. Only a powerful state, a growing number of commentators and critics suggested, could arrest or perhaps reverse these growing inequalities; only the state could guarantee the individual's well-being. 'I hold it for indisputable', wrote Ruskin as early as 1853 (*Stones of Venice*, Volume III),

> that the first duty of a State is to see that every child born therein shall be well housed, clothed, fed, and educated, till it attain years of discretion. But in order to the effecting this the Government must have an authority over the people of which we now do not so much as dream.
>
> (Cook and Wedderburn 1903–12, Vol. XI, 263)

Increasingly, society's critics leant towards more collectivist forms of political thinking, thinking that even extended to socialism (Briggs 1959 [1974], 474–5). Ruskin himself was an inspiration. By the start of the new century, remarked the admiring E. T. Cook, Ruskin's heresies were becoming 'accepted doctrine' (1912, 11).[3]

To many, including Arnold, the state provided the means of addressing the gross inequalities created by unfettered capitalism: state intervention and regulation could and would counter-act its pernicious, divisive, and destructive influence.[4] This was the path that many Western societies pursued in the twentieth century, particularly in the immediate post-war era of Bretton Woods, when a Keynesian form of embedded liberalism was in the ascendant (Harvey 2005 [2007], 10–11; Gould 2013, 6).[5] Yet this reformed capitalism did not long survive (Harvey 2005 [2007], 12–16). As economic growth began to stall in the 1970s, there were growing calls to restore a free market, shrink the state, and enable the sovereign individual to make her or his own way in society, unfettered by rules and regulations (Gould 2013, 4–5). The Bretton Woods agreements finally collapsed in 1973, after the so-called Nixon Shock of 1971, when America effectively withdrew its pivotal support for the arrangement, in large part – and somewhat ironically – because of its own spending on a conflict by definition at odds with the peaceable intentions of Bretton Woods. Attention returned to the individualistic ethos and *laissez faire* liberal economics that dominated early Victorian society. Famously, Milton Friedman captured this shift in thinking, and gave it conceptual definition (Gould 2013, 5). The free market and free enterprise were the essential underpinnings of a free society, Friedman wrote in his landmark text, *Capitalism and Freedom*, 'a necessary condition for political freedom' (1962 [2020], 6); anything that interfered with the freedom of the marketplace was the enemy, by extension, of freedom itself. As he added in a piece first published in the *New York Times* on 13 September, 1970, business had no responsibility to do anything but maximise its own profits.

> [Yet] businessmen believe that they are defending free enterprise when they declaim that business is not concerned 'merely' with profit but also

> with promoting desirable 'social' ends; that business has a 'social conscience' and takes seriously its responsibilities for providing employment, eliminating discrimination, avoiding pollution and whatever else may be the catchwords of the contemporary crop of reformers. In fact they are – or would be if they or anyone else took them seriously – preaching pure and unadulterated socialism. Businessmen who talk this way are unwitting puppets of the intellectual forces that have been undermining the basis of a free society these past decades.
>
> (Friedman 2007, 173)

This new form of liberalism, or neoliberalism, swiftly came to dominate the political landscape in the United Kingdom and the United States (Stiglitz 2002, 13), drawing widespread assent from across the political divide – and elsewhere in Western societies (Harvey 2005 [2007], 9; Gould 2013, 7). In time, neoliberalism became an integral part of the developed world's way of thinking: large sections of society took it for granted that human freedom and human flourishing were alike dependent on continued economic growth, which demanded a global free market, unfettered by nation-state intervention, and unburdened by the kind of taxation that might be used to even out the inequalities the free market continued to create and sustain.[6]

The rise to dominance of this 'free-market ideology' (Stiglitz 2002, 13) was assisted by another historical development. As we have seen, Victorian society pursued reformist policies to address the problem of inequality, leading with extensions to the political franchise that offered previously marginalised sections of the community a greater say in the role and running of society and state. But to the more radically-minded, reforming the system was a futile exercise; capitalism was by its very nature a system that created – so it could then sustain and exploit – inequality. To Karl Marx and Friedrich Engels, authors of *The Communist Manifesto* (1848), the only way in which to achieve a truly just society was through revolutionary change, which would in turn lead to a post-capitalist society.

Socialism was not Marx's creation, but his interpretation of it became perhaps the single most powerful force for change in twentieth-century history, sparking revolutions around the world. By that century's end, however, the great Marxist experiment was itself faltering (Claeys 2018, 445). Ostensibly socialist states such as Soviet Russia had failed to deliver on the promise of material prosperity, whilst simultaneously depriving their citizens of the freedoms and liberties enjoyed in Western liberal democracies (Piketty 2021, 128; Claeys 2018, 445–6). Despite the many ways in which these socialist states failed to live up the post-capitalist ideal proposed by Marx, it appeared to enemies and allies alike that Marxism had itself failed (Gould 2013, 3). Without a socialist or communist 'counter-model' to capitalism (Piketty 2021, 128), some argued that history itself had come to an end.[7] There was nothing now in humankind's future but the endless extension of the free market, and the continuing multiplication of the liberal democracies that pioneers of neoliberalism such as Friedman insisted were the inevitable concomitant of capitalism itself (Gould 2013, 1–2).

Yet these predictions were (as predictions so often are) proven wrong. In communist China, for example, the first glimmers of a counter-revolution were immediately and ruthlessly extinguished. In its wake, a totalitarian state concluded that capitalism was precisely what the party needed. China is now one of the most powerful economies in the world, yet it remains in all its leading particulars an authoritarian one-party state (Piketty 2021, 226). Capitalism and freedom, it transpired, were not after all eternally aligned. Nevertheless, the collapse of Soviet Russia – and the widespread sense that Marxism itself had been discredited – did have an influence on the emerging field of environmental justice studies (or EJ Studies), then nascent in the United States. For good reasons, leading scholars such as Robert D. Bullard focused on the role that race played in environmental injustice. In so doing, however, environmental justice scholars drew back from inquiring too deeply into the other causes of environmental injustice, not least the role that capitalism itself played in perpetuating inequality (Fraser 1997, 1–2).[8] Class had lost its explanatory power; identity politics took its place, as itself a fruitful way in which to approach the problem of environmental injustice, since questions of gender and race were also factors (Fraser 1997, 1–2). Indeed, US environmental justice scholarship frequently conflates environmental injustice with environmental racism. Nevertheless, and as its critics have pointed out, this failure to engage with the relationship between capitalism and environmental injustice has, in the context of a return to the kind of globalised free market that the Victorians themselves pioneered, significantly hampered the effectiveness of the environmental justice movement in tackling environmental injustice (Bell 2014, 16; Pulido 2018, 17, 21). The problem, as left-leaning scholars have pointed out, is that capitalism – or capitalism in its unregulated, free-market form – has come to seem so inevitable (and even 'natural') a part of the political landscape that imagining any alternative is, even today, difficult, if not impossible (Harvey 2005 [2007], 3; Gould 2013, 5–8; Monbiot 2016 [2017], 3).

In a strange twist, therefore, the economic and political structures of twenty-first-century societies are dominated by doctrines that would have been more-or-less familiar to the early Victorians: late-modern societies have been returned, in other words, to a mind-set unmodified by the struggles of the years that followed. It is said that those who forget their history are condemned to repeat it (a paraphrase of George Santayana; 1905 [1910], 284). At a moment in human history when most people now live in cities, the United Nations Human Settlements Programme (2018) calculates that 29% of the world's urban population still lives in slums.[9] What constitutes a slum, and its converse? The United Nations Human Settlement Programme (UN-Habitat) uses multiple criteria to define a slum, including its location 'in dangerous or polluted sites or in immediate proximity to pollution sources' (UN-Habitat 2018, 13). Yet the problem of poverty and pollution, as the

United Nation's Technical Report on Environmental Justice (UNDP 2022) insists, has only been compounded by wider failures to relate and respond to social deprivation *and* environmental degradation:

> human action – and inhuman inaction – have brought us to a "triple planetary crisis" of climate change, pollution and nature loss that has a direct and severe impact on a broad range of human rights, including the rights to adequate food, water, education, housing, health, development, and even life itself.
>
> (UNDP 2022, 8)

As the UN's Report on Environmental Justice continues, these

> interlinked planetary crises act as threat multipliers, amplifying conflicts, tensions and structural inequalities, and forcing people into increasingly vulnerable situations … The COVID-19 pandemic further aggravates this scenario, exposing deep inequalities in the distribution of wealth and resources, in the delivery of basic services, in the promotion of justice and security, and in the protection of human rights. Extreme poverty is rising for the first time in two decades.
>
> (UNDP 2022, 9)

In spite, therefore, of those who still believe that wealth will trickle down to the disadvantaged (libertarian, free-market fundamentalists), or that wealthy societies are those with both the will and wherewithal to tackle environmental questions (ever hopeful apologists, fingers pointing to an imaginary environmental Kuznets curve), neoliberalism has done little to curb global inequality, exploitation, and environmental depredation. To the contrary, the economic growth it has fostered has magnified inequality and worsened environmental injustice – whether it is conceived of in narrowly anthropocentric terms, or in the more expansive sense of an emergent justice. Should nations such as the United Kingdom therefore unwind the years, and return to the collectivist, state-led policies of the mid-twentieth century? The post-war period itself marked what has been called the Great Acceleration (Steffen et al. 2015), as human impacts on the nonhuman world increased exponentially; some identify this as the start of the Anthropocene (McNeill and Engelke 2014). Effective as state intervention may have been in addressing human inequalities – and the evidence suggests it was – the interventionist approach was still tied to the paradigms of industrial growth, capital accumulation, and progressivist modernisation.[10] In fact, it has been argued that the state is itself an active contributor to environmental injustice (Pellow 2018, 22–4). Can a state-based approach to these fundamental difficulties of environmental and emergent justice be refashioned in ways that would more fully recognise the constitutive importance of the nonhuman and natural worlds? Certainly, it would seem that the state alone has the power to arrest the atomisation of society, tackle its interlinked inequalities, and promote wealth-as-life, intervening between 'nature as tap and sink and capital accumulation' (Moore 2016, 10); just as certainly, 'we are likely stuck with states for the time being' (Pellow 2018, 24).

Before (re)considering the concept of the strong state, however, we might perhaps reconsider the views of one of the most notable of Victorians discussed in this book: Ruskin himself. Ruskin would have been surprised by the emphasis that has sometimes been placed on his ideas. Ruskin did not believe in equality (Cockram 2007, 43). He believed that all people were born unequal, just as humankind itself occupied a position within an orderly, hierarchical world that set it above other species. Relationality and co-dependency did not therefore mean that value was distributed equally, only that it was distributed across the species divide, in ways which the prudent would do well to respect. What mattered, in his estimation, was equality of opportunity – an equal chance of finding ways in which to flourish – and since flourishing was (beyond what might be called the bare necessities of existence) a personal matter, it followed that the state could not in the final analysis decide what it should involve. That last step towards 'true felicity' of circumstance and well-being lay in the hands of the individual (Cook and Wedderburn 1903–12, Vol. XVII, 111). 'Note, finally', Ruskin wrote in *Unto this Last*, 'that all effectual advancement towards this true felicity of the human race must be by individual, not public effort. Certain general measures may aid, certain revised laws guide, such advancement; but the measure and law which have first to be determined are those of each man's home' (Cook and Wedderburn 1903–12, Vol. XVII, 111–2).

It was a point that, amongst those influenced by Ruskin, William Morris (1834–96) perhaps grasped most fully. Variously an artist, printer, designer, translator, poet, conservationist, and activist, Morris (2003 [2009]) wrote one of the most important – but also idiosyncratic – responses to industrial modernity, the utopian fantasy, *News from Nowhere* (1890). The novel's idiosyncrasies derive, in part, from Morris's fusion of a Ruskinian emphasis on life as the only meaningful form of wealth, a progressivist Marxist belief in a historical dialectic which would, through revolution, give way to a classless future, and Morris's instinctual contempt for industrial modernity and its products or effects, both intentional and unintentional. But the novel also owes something to the anarchist visions of Morris's acquaintance Peter Kropotkin (Geus 1999, 105), whose mindfulness of human and nonhuman interdependencies and arguments for 'a libertarian ecological structuring of society' (102) anticipate today's bioregionalist arguments (102). It is from Kropotkin, notes Marius de Geus, that Morris takes his emphasis on 'finding room for freedom' (1999, 102), and in particular, 'the freedom to do what one is best at' (108).

That emphasis anticipates the capabilities approach (discussed in Chapter 5) pioneered by Amartya Sen, for whom 'social justice' is practically embodied in 'the capabilities that a person has, that is, the substantive freedoms he or she enjoys to lead the kind of life he or she has reason to value' (Sen 1999, 87). Morris himself believed that a just world could not be achieved without attending to all the conditions necessary to a life lived fully and wholly: in the post-revolutionary future that he envisaged in *News from Nowhere*, society has invested in 'the whole complex of human fulfilments' (Wilmer 1993 [1998], xxxxvii). Utility (and use-value) are not the only measure of what constitutes the good life and a just society. In the aftermath of revolution, as Morris's time-travelling narrator discovers, 'a craving for beauty seemed to awaken in men's minds' (Morris 2003 [2009], 115). The citizens of *Nowhere* are, he writes, a 'beauty-loving people' (Morris 2003 [2009], 30): they set out to engender beauty, and 'live amidst beauty' (62). Beauty has become an operative principle in shaping material relationalities of every kind, signifying 'a creative turn to the sensuous body' (Eagleton 1990, 9). The result is what one of Morris's characters describes as a 'delight in the life of the world; intense and overweening love of the very skin and surface of the earth on which man dwells' (Morris 2003 [2009], 113).

In the future that Morris envisages, this focus on the aesthetic dimension of lived existence enacts a decisive shift in relationalities of all kinds, including those that shape the land itself, now seen for what it is – a naturalcultural hybrid – and worked with 'regard to beauty' as well as utility, 'everywhere treated as a garden made for the pleasure as well as the livelihood of all' (Morris 2003 [2009], 164). This consideration extends to all kinds of naturalcultural assemblage, from orchard to forest and 'waste' (Morris 2003 [2009], 63–4), some spaces cultivated, some simply enjoyed, others left alone, but all regarded in light of a deep responsiveness to '[t]he earth and the growth of it and the life of it' (174). Here, justice emerges, tentatively, as a form of respect that extends beyond the human, and as the natural expression of a society that has (in a turn of phrase that echoes Ruskin) 'cast away riches and attained to wealth' (Morris 2003 [2009], 172).

Morris's vision of the future – at once a 'bold escape from modernity' (Bonnett 2010, 70) and a compelling 'site of counter-orthodoxy' (71) – is, therefore, less the expression of abstract principle or totalising vision than the practical consequence of a way of life that hinges on a 'palpable apprehension of the material interweaving of human and more-than-human worlds' (Marland 2015, 126), and of the complex consequences of that entanglement. It is a society governed by 'Equality of Life' (Morris 2003 [2009], 165), where equality is not understood as sameness, but as (bio)diversity, whose purpose is to engender an 'abundance of life' (21). What this requires, and what Morris proposes, is a loose, anarcho-socialist framework of decentralised and cooperative governance; what he proposes, to return to David Pellow's point, is the building of 'environmentally just and resilient communities *beyond* the state' (Pellow 2018, 22; see also Piketty, who defends a 'democratic, decentralised socialism' as an effective path to equality; 2021, 226).

If, now, we glance back to Dickens's own response to the conditions created by industrial modernity – the circus in *Hard Times*, discussed in Chapter 5 – we can see 'Sleary's Horse-riding' (Dickens 2006 [2008], 257) as itself a kind of self-sustaining yet anarchic assemblage that exists out of reach of the state or its dominant paradigms. But perhaps the most interesting Victorian illustration of the argument for an anarchic, sustainable community can be found in the work of Richard Jefferies (1848–87). Morris was deeply affected by Jefferies' *After London* (1885), 'a book that Morris afterwards was never weary of praising' (Mackail 1899, 144), but Jefferies's final novel, *Amaryllis at the Fair* (1887), is an equally plausible influence on *News from Nowhere*. Jefferies's *Amaryllis* is itself, in part, an acerbic commentary on the growing inequalities of Victorian society – on the segregation implicit in a now industrialised society, and the willed ignorance of the rich – but it juxtaposes that commentary with the idyll of Coombe Oaks, a small-holding modelled on the farm where Jefferies grew up.[11] Here, the novel's central character, Iden, conspicuously fails to transform his hard work into a successful business. Nonetheless, he does succeed in creating a flourishing antidote to the ecological exile of a denatured, urban existence. Coombe Oaks is a 'spa among the apple-bloom' (Jefferies 1887, 229–30), where the ale (Iden's own) is a 'potable sunbeam' (230), and there are 'crumbs for everybody, and for the robin too' (251). It is an equitable place, in which the farm, its visitors, and its workers all come together in easy association, free to share ideas (on love, on politics, on life), in a kind of *respublica literaria*. But the most noteworthy aspect of this republic lies in Iden's ability to read the nonhuman world of nature as if it was itself lettered; 'for Iden could talk of the trees and grass, all that the Earth bears, as if one had conversed face to face with the great god Pan himself' (Jefferies 1887, 257). The result is a kind of natural prodigality:

> Flowers, and trees, and grass, seemed to spring up wherever Iden set down his foot: fruit and flowers fell from the air down upon him. It was his genius to make things grow – like sunshine and shower … a sort of Nature in human shape, moving about and sowing Plenty and Beauty.
>
> (Jefferies 1887, 199)

Environmental injustice assumes a degraded, dangerous environment whose hazards impact on the impoverished and marginal; its corollary is a place where the environment itself flourishes, in 'Plenty and Beauty', and its flourishing makes possible a whole and healthy human life, justly available to all. Flawed as it may be, fraught as life there sometimes is, Iden's Coombe Oaks embodies that sense of what a just society might look like.

'Such a mixture never was', wrote the poet Edward Thomas in his commentary on the novel (1909, 285) – and perhaps never shall be, as Jefferies himself acknowledged. Even as the reader is left in the happy company of Iden and his fellow creatures, Jefferies insists that 'their Interlude in Heaven' (1887, 260) amounts to nothing more than a brief interruption in the flow of historical inevitability. 'Nothing ever happens in the fitness of things', observes the novel's narrator (1887, 259), to which one might add the words of Dickens's character R. Wilfer, quoted in the Introduction: that 'what might have been is not what is!' (Dickens 1864–5 [1997], 42). Then as now, responses such as Morris's utopian imagining or Jefferies's pastoral idyll beg obvious questions. Are visions such as these practicable? Can they ever be transposed to our own world? The modern equivalent of these ideas is, perhaps, the concept of the bioregion, a 'biotically determined' alternative to state-based modes of governance (Lynch, Glotfelty, and Armbruster 2012, 2), grounded in a 'reinhabitory discourse' (2012, 21) itself characterised by a '[d]efiant decentralism' (3); as its advocates insist, the concept of the bioregion represents an equitable, sustainable form of living attuned to naturalcultural hybridity.[12] It remains to be seen whether a bioregional form of existence can survive the now globalised architecture that capitalism has engendered, but according to green economists such as Molly Scott Cato, it not only can, but must: 'the costs of global competition are too high' (Cato 2013, 207). Moreover, Cato adds, the origins of today's modern market economy in Britain's own economic history 'places on its academics and intellectuals a peculiar authority matched by a particular responsibility' (2013, 218).

As this discussion underlines, the problems identified by an earlier generation of Victorians – and explored throughout this book – continued to occupy their successors later in the century, generating fresh ideas about how best to address the iniquitous world that was industrial modernity's apparently inevitable corollary. By the time Wells, White, Morris and Jefferies were writing, however, the transformation of society seemed irreversible: for Britain, writes Boyd Hilton, 'there could be no going back' (2006, 629), and in the race to modernise, 'other nations were hard on her heels' (630). Herein lies the particular value of the generation of writers discussed in this book. In the first decades of Victoria's reign, the conditions created by industrial modernity were still new, and reform had yet to mitigate its most obvious effects. Many could remember what the world was like before the horizon was crowded with chimneys, and towns and cities sharply segregated into suburbs and slums, factory quarters, and nuisance neighbourhoods; they had not yet been persuaded that this world was inevitable or somehow natural. Throughout these years – and particularly during the decade known as the Hungry Forties – critics and commentators responded with often incendiary indictments of the changed circumstances of industrial existence, and their impact on ordinary people, ordinary lives, and ordinary life-worlds. What was done could yet be undone; and seized by a sense of moral indignation at the human costs of change – and by nascent concerns about the deeper impact of ecological degradation – the writers of the period confronted what was happening around them, and posited alternatives to it. Today, and at a moment when many (particularly in the developed world) seem unable to conceive of an alternative to a political and economic landscape dominated by neoliberalism, their literary interventions take on a fresh importance.

*

During the nineteenth century, Britain led the world in a transformation that had dramatic consequences for humankind's impact on the nonhuman world. The nature of those effects was not then fully appreciated; closer to home, however, the convergence of free-market capitalism and idustrialisation created recognisably modern forms of environmental injustice and inequality, centred on class. Particularly during the second quarter of the nineteenth century, the urban environment deteriorated dramatically, both because of rapid industrialisation, and the urbanisation to which it gave rise. The result was an emergent risk society, in which industrialisation and urbanisation generated new and often unseen threats to health and well-being. This process was always uneven. As Ruskin wrote in *Unto this Last*, 'riches are a power … acting only through inequalities or negations of itself' (Cook and Wedderburn 1903–12, Vol. XVII, 44). The rich were not affected in the same way as the poor, and the gulf between rich and poor took on a spatial form, as the better-off retired to the safety of the suburbs, leaving the poor to make the best of the slums, or negotiate new risks in factory neighbourhoods. Inequality was environmental: it was written into the world in which the poor lived and worked, and it shaped their very bodies. Even as society professed its own ignorance of this environmental classism, however, critics and commentators found ways in which to challenge the Victorian faith in the form of 'progress' that society celebrated. In works of prose, poetry, and fiction, writers from both working-class and middle-class backgrounds interrogated the kind of society that was being built around them, critiquing its impact on the marginal and disadvantaged, whilst asking profound and important questions about its wider implications for the constitutive entanglements that knit people and place, human and nonhuman. In so doing, these early Victorian writers began to develop a notion of a fair and just society that overlaps not only with today's environmental justice perspectives, but with modern, new materialist conceptions of ethical accountability and emergent justice.

*

Notes

1 In Preston, for example, the rich set themselves up in Winckley Square, in the southwest of the town (Morgan 1995, Ch. 1). In Birmingham, they moved to Egbaston (Steinbach 2017, 20). In Manchester, wrote Engels, the 'upper bourgeoisie' established themselves 'on the breezy heights of Cheetham Hill, Broughton, and Pendleton, in free, wholesome country air, in fine comfortable homes' (1993 [2009], 58). Once established, this pattern was everywhere emulated, quickly extending to those who, like R. Wilfer in Dickens's *Our Mutual Friend*, could only just affords the means to situate themselves as suburbanites. By the century's end, the problem was not so much the exclusivity of the suburban mode, but its apparent ubiquity and uniformity – always more imagined than real, as Sarah Bilston points out (2019, 5–10). More to the point, the suburban way of life was

still socially exclusive, ruling out the large majority of the working classes and all those identified as society's residuum.

2 Booth wrote with his namesake, General William Booth, in mind: in 1890, General Booth famously argued that the Victorian city was not only as remote and unknown as Africa, but as dark and dangerous (Booth 1890, 11–12). With a nod to the explorer Henry M. Stanley's *In Darkest Africa* (1890), General Booth called his exposé *In Darkest England*, arguing that, '[a]s there is a darkest Africa is there not also a darkest England? … May we not find a parallel at our own doors, and discover within a stone's throw of our cathedrals and palaces similar horrors to those which Stanley has found existing in the great Equatorial forest?' (Booth 1890, 11–12).

3 When in 1906 the newly established Labour Party won 29 seats in Parliament, the journalist W. T. Stead canvassed the new members to find out which writers had been the greatest influence upon them: Ruskin was the favourite (Cook 1912, 14).

4 The related question, as P. J. Keating highlights in his analysis of Matthew Arnold's social and political thinking, is whether this emphasis on a strong state itself amounts to a form of authoritarianism; as Keating argues, this depends entirely on one's own political persuasions (1975, 207–8). In the immediate post-war, post-1945 period, many western democracies developed strong states, in part to avoid the economic crises that, in the 1930s, allowed totalitarianism to prosper in Europe, and in part to ensure a new level of social equality. To some, both then and since, this development would itself have seemed to entail a denial of freedoms; it certainly interfered with the free play of industrial capitalism.

5 Put in place in 1944, the Bretton Woods accords created a binding, international system of monetary cooperation that laid the basis for a prolonged period of post-war peace and prosperity. It also created institutions such as the International Bank for Reconstruction and Development, better known as the World Bank, whose purpose was to assist post-war reconstruction and 'save the world from future economic depressions' (Stiglitz 2002, 11). More recently, notes Joseph Stiglitz, the World Bank has become an important force in driving a free-market agenda itself at odds with the Bretton Woods agenda (2002, 13), creating rather than averting economic crises (15).

6 As David Harvey points out, 'increasing social inequality' has been 'such a persistent feature of neoliberalisation are to be regarded as structural to the whole project' (2005 [2007], 16). In *Violence in Capitalism* (2016), James A. Tyner takes the point further, arguing that the 'market logics' of capitalism ('but especially the neoliberal, neoconservative variant') 'are determinant of a pervasive indifference to life whereby some individuals are disallowed life because they fail to conform to the dictates of capital accumulation'; in this discussion of capitalism's inherent, structural violence, laissez faire becomes '*a market logic of … "letting die"*' (italics in the original; Tyner 2016, 10).

7 It was Francis Fukuyama who famously argued for the end of history (1992 [2020]), and it is Fukuyama's argument that the economist Stephen D. King invokes in *Grave New World: The End of Globalization and the Return of History* (2017). Like Stiglitz (2002), King critiques globalisation, which has 'simply failed to deliver prosperity to all' (2017, 12) but what he in fact describes is the extensification of capitalism through new forms of technology (1), about which there is nothing new. Capitalism is by its nature a global project, with global aspirations, the point made by Engels and Marx in *The Communist Manifesto* (2002, 223) and quoted in Chapter 1.

8 In the United States, any form of anti-capitalist argument is and always has been controversial. Given the inherent conservatism of the US political landscape, noted by Joseph A. Schumpeter (1943 [2010], 3), it is not surprising that socialism should

be seen as a form of heresy. Moreover, and as Bryan Gould notes, socialism is in the US routinely conflated with communism and Marxism, between and within which there are in fact profound differences; for example, Marx's 'Economic and Philosophical Manuscripts' (1844) – on which I touched in Chapter 2 – were for many years 'regarded as so subversive' of Marxist orthodoxy that they were left unpublished (Claeys 2018, 59). With Soviet Russia's collapse, it may be argued that socialism then lost what hold it still had on thinkers and activists alike. Certainly, the first-wave of US-centric environmental justice scholarship was averse to pursuing any argument that seemed to indict free-market capitalism as a cause of environmental injustice, no matter how compelling the evidence.

9 The Population Division of the UN Department of Economic and Social Affairs (UN DESA) estimates (2018) that 55% of the world's population now lives in urban areas, a figure expected to increase to 68% by 2050.

10 Walter Scheidel refers to 'the great compression' of inequality (Scheidel 2017 [2018], pp. 170–1), Thomas Piketty to the 'great redistribution' of wealth (2021, 121); both see it as a process prompted by the two world wars, which continued into the post-war decades. This reduction of inequality is in marked contrast to the more recent 'great divergence', as the Nobel laureate Paul Krugman has described the period since the 1970s (2007 [2009], 124); see also Timothy Noah's *The Great Divergence* (2012). Although both Krugman and Noah focus on the US-American experience of growing inequality, the work of historians such as Piketty (2021, 154) highlights the extent to which this pattern has been reproduced across other Western societies. Piketty nevertheless finds grounds for hope in the fact that 'inequalities of income [are nevertheless] appreciably smaller in 2020 than in 1910' (2021, 154), adding that '[t]o continue the march towards equality, the most natural path seems to be already blazed' (155). There is much at stake in this discussion. Despite – or perhaps because of – Piketty's measured approach to his subject matter, his arguments have prompted a backlash that recapitulates all the standard arguments for neoliberalism, whilst insisting that Piketty simply enables 'the left to persist in its errors since communism collapsed and socialism failed everywhere' (Delsol and Martin 2015 [2017], n.p.).

11 The poor are safely contained within the slums, wrote Jefferies; the rich have retreated to the suburbs and beyond; 'round about this great Babylon of Misery', says the narrator, apropos London, 'the rich folk ... dwell in ease and luxury at Sydenham, at Norwood, at Surbiton, at Streatham, at Brighton, at Sevenoaks, wherever there is pure air' (Jefferies 1887, 223–4). Their idea of charity, adds the narrator, is to build 'Homes for Lost Dogs and Homes for Lost Cats, neither of which are such nuisances as human beings' (1887, 224). Like the poor in White's short story, however, the poor refuse to stay where they are put. These 'human nuisances' (1887, 225) spill out onto fashionable streets like the Strand. It is one thing to ignore what it out of sight; another to ignore what is in plain sight. Yet the rich somehow contrive it: starving children 'play on the pavement unregarded. Hatless, shoeless, bound about with rags, their faces white and scarred with nameless disease, their eyes bleared, their hair dirty ... How *can* people pass without seeing them?' (italics in the original; 1887, 223).

12 Marius de Geus draws a distinction between the modern 'utopia of abundance', where happiness is equated with materialism, and an ecologically responsible 'utopia of sufficiency', like Morris's, which flows from 'simplicity, self-restraint and moderation' (1999, 21). There have been several attempts to translate this principle of sufficiency into a working model for society, but the bioregion is one of the better known; originating on the West Coast of America in the 1970s, it was later developed by Kirkpatrick Sale, 'the leading ideologue of bioregionalism' (Cato 2013, 38). The 'bioregion' is an area in which ecological footprint and

geographical limits are aligned, each supporting the other, and personal and communal identity are closely related to place itself. 'To become dwellers in the land', Sale wrote, 'to come to know the earth, fully and honestly, the crucial and perhaps only and all-encompassing task is to understand the place, the immediate, specific place, where we live' (1984 [1986], 224).

Works cited

Arnold, Matthew. 1965. *The Complete Prose Works, Volume V*, edited by R. H. Super. Ann Arbor, MI: University of Michigan Press. Available at archive.org, https://archive.org/details/completeprosewor0005arno/page/n5/mode/2up. Accessed 2 November 2022.

Beck, Ulrich. 1992 [2012]. *Risk Society: Towards a New Modernity*, translated by Mark Ritter. London: Sage.

Bédarida, François. 1970 [1985]. 'Cities: Population and the Urban Explosion'. *The Nineteenth Century: The Contradictions of Progress*, edited by Asa Briggs, 95–118. London: Thames & Hudson.

Bell, Karen. 2014. *Achieving Environmental Justice: A Cross-National Analysis*. Bristol: Policy Press.

Bilston, Sarah. 2019. *The Promise of the Suburbs: A Victorian History in Literature and Culture*. London: Yale University Press.

Bonnett, Alastair. 2010. *Left in the Past: Radicalism and the Politics of Nostalgia*. London: Continuum.

Booth, Charles. 1903. *Life and Labour of the People in* London – *Second Series: Industry*, I. London: Macmillan. Available at archive.org, https://archive.org/details/lifelabourofpeop01boot. Accessed 9 November 2021.

Booth, General (William). 1890. *In Darkest England and the Way Out*. London: International Headquarters of the Salvation Army. Available at archive.org, https://archive.org/details/a548464700bootuoft/page/n9/mode/2up?view=theater. Accessed 8 November 2021.

Brattin, Joel J. 2002. 'Constancy, Change, and the Dust Mounds of *Our Mutual Friend*'. *Dickens Quarterly* 19 (1): 23–30. Available at jstor.org, www.jstor.org/stable/45291848. Accessed 16 October 2022.

Briggs, Asa. 1959 [1974]. *The Age of Improvement, 1783–1867*. London: Longman.

Carlyle, Thomas. 1843 [1897]. *Past and Present. The Works of Thomas Carlyle, Volume Six*, 417–500. New York, NY: Peter Fenelon Collier. Available at archive.org, https://archive.org/details/worksofthomascar0006/page/n5/mode/2up?view=theater. Accessed 8 December 2022.

Cato, Molly Scott. 2013. *The Bioregional Economy: Land, Liberty and the Pursuit of Happiness*. London: Routledge.

Claeys, Gregory. 2018. *Marx and Marxism*. London: Pelican.

Cockram, Gill G. 2007. *Ruskin and Social Reform: Ethics and Economics in the Victorian Age*. London: Tauris Academic Studies.

Cook, E. T. 1912. *The Life of John Ruskin, Vol. II*: 1860–1900. 2nd edn. London: George Allen & Company.

Cook, E. T., and Alexander Wedderburn (eds). 1903–12. *The Library Edition of John Ruskin's Works*. 39 vols. London: George Allen. Available at www.lancaster.ac.uk/the-ruskin/the-complete-works-of-ruskin/. Accessed 5 January 2023.

Cubitt, Eliza. 2019. *Arthur Morrison and the East End: The Legacy of Slum Fictions*. London: Routledge.

Delsol, Jean-Philippe, and Emmanuel Martin. 2015 [2017]. 'Introduction'. *Anti-Piketty: Capital for the 21st-Century*, edited by Jean-Philippe Delsol Nicolas Lecaussin, Emmanuel Martin n.p. Washington, DC: Cato Institute.

Dickens, Charles. 2006 [2008]. *Hard Times*, edited and introduced by Paul Schlicke. Oxford: Oxford University Press. First published 1854.

Dickens, Charles. 1864–5 [1997]. *Our Mutual Friend*, edited and introduced by Adrian Poole. London: Penguin.

Disraeli, Benjamin. 1981 [2008]. *Sybil; or, The Two Nations*, edited by Sheila M. Smith. Oxford: Oxford University Press. First published 1845.

Eagleton, Terry. 1990. *The Ideology of the Aesthetic*. Oxford: Basil Blackwell.

Engels, Friedrich. 1993 [2009]. *The Condition of the Working Class in England*, edited and introduced by David McLellan. Oxford: Oxford University Press.

Fraser, Nancy. 1997. *Justice Interruptus: Critical Reflections on the 'Postsocialist' Condition*. London: Routledge.

Friedman, Milton. 1962 [2020]. *Capitalism and Freedom*. Chicago, IL: University of Chicago Press.

Friedman, Milton. 2007. 'The Social Responsibility of Business Is to Increase Its Profits'. *Corporate Ethics and Corporate Governance*, edited by Walther C. Zimmerli, Klaus Richter, and Markus Holzinger, 173–8. Berlin: Springer.

Fukuyama, Francis. 1992 [2020]. *The End of History and the Last Man*. London: Penguin.

Geus, Marius de. 1999. *Ecological Utopias: Envisioning the Sustainable Society*. Utrecht: International Books.

Greenwood, James. 1873. *In Strange Company: Being the experiences of a roving correspondent*. London: Henry S. King & Co. Available at archive.org, https://archive.org/details/instarangecompa00greegoog/page/n6/mode/2up. Accessed 3 November 2021.

Gould, Bryan. 2013. *Myths, Politicians and Money: The Truth behind the Free Market*. Houndmills: Palgrave Macmillan.

Hall, Peter. 2002 [2003]. *Cities of Tomorrow*. 3rd edn. Oxford: Blackwell.

Harvey, David. 2005 [2007]. *A Brief History of Neoliberalism*. Oxford: Oxford University Press.

Hilton, Boyd. 2006. *A Mad, Bad, and Dangerous People? England 1783–1846*. Oxford: Clarendon Press.

Horn, Pamela. 1997 [1999]. *The Victorian Town Child*. Stroud: Sutton.

James, Henry. 1893. *Essays in London and Elsewhere*. New York, NY: Harper & Brothers. Available at archive.org, https://archive.org/details/essaysinlondone l00jame. Accessed 9 November 2021.

Jefferies, Richard. 1887. *Amaryllis at the Fair: A Novel*. London: Sampson Low, Marston, Searle, & Rivington. Available at archive.org, https://archive.org/details/amaryllisatfair02jeffgoog/page/n6/mode/2up?view=theater. Accessed 6 May 2022.

Jefferies, Richard. 1907 [1913]. *The Story of My Heart: My Autobiography*. London: Longman, Green, and Co. First published 1883.

Jones, Ernest. 1852. 'The Social War: Kind Masters, Slavish Men'. The People's Paper: *The Champion of Political Justice and Universal Right*, 13 November. Available at the British Newspaper Archive, www.britishnewspaperarchive.co.uk

/viewer/BL/0002582/18521113/007/0001?browse=true. Accessed 2 November, 2021.

Keating, P. J. 1975. 'Arnold's Social and Political Thought'. *Writers and their Background: Matthew Arnold*, edited by Kenneth Allott, 207–35. London: G. Bell & Sons.

King, Stephen D. 2017. *Grave New World: The End of Globalization and the Return of History*. London: Yale University Press.

Krugman, Paul. 2007 [2009]. *The Conscience of a Liberal: Reclaiming America from the Right*. London: Penguin.

Liebreich, Karen, and Sarah Cruz. 2018. *The Chiswick Timeline: A History in Art and Maps*. London: McHugh Publications for Abundance London.

Luckin, Bill. 2015. *Death and Survival in Urban Britain: Disease, Pollution and Environment 1800–1950*. London: Bloomsbury Academic.

Lynch, Tom, Cheryll Glotfelty, and Karla Armbruster. 2012. 'Introduction'. *The Bioregional Imagination: Literature, Ecology and Place*, edited by Tom Lynch, Cheryll Glotfelty, and Karla Armbruster, 1–30. Athens, GA: University of Georgia Press.

Mackail, J. W. 1899. *The Life of William Morris, Volume II*. London: Longmans, Green & Co. Available at archive.org, https://archive.org/details/lifeofwilliammo r02mackuoft/page/144/mode/2up?q=weary. Accessed 22 July 2020.

Marland, Pippa. 2015. 'The Gannet's Skull Versus the Plastic Doll's Head: Material 'Value' in Kathleen Jamie's 'Findings''. *Green Letters* 19 (2): 121–31. doi: https://doi.org/10.1080/14688417.2015.1024156

Marx, Karl, and Friedrich Engels. 1848 [2002]. *The Communist Manifesto*, introduced by Gareth Stedman Jones. London: Penguin.

McNeill, J. R., and Peter Engelke. 2014. *The Great Acceleration: An Environmental History of the Anthropocene since 1945*. Cambridge, MA: Belknap Press.

Monbiot, George. 2016 [2017]. *How Did We Get Into This Mess? Politics, Equality, Nature*. London: Verso.

Moore, Jason W. 2016. 'Introduction'. *Anthropocene of Capitalocene? Nature, History, and the Crisis of Capitalism*, edited by Jason W. Moore, 1–11. Oakland, CA: PM Press.

Morgan, Nigel. 1995. *Desirable Dwellings: Middle Class Housing in Preston in the First Half of the 19th Century*. Unpublished manuscript. Available at prestonhistory .com, https://prestonhistory.com/subjects/desirable-dwellings-chapter-one-choice -locations/. Accessed 30 October 2021.

Morris, William. 2003 [2009]. *News from Nowhere, or, An Epoch of Rest, Being Some Characters from a Utopian Romance*, edited by David Leopold. Oxford: Oxford University Press. First published 1890.

Noah, Timothy. 2012. *The Great Divergence*. London: Bloomsbury.

Pellow, David Naguib. 2018. *What is Critical Environmental Justice?* Cambridge: Polity.

Piketty, Thomas. 2021. *A Brief History of Equality*, translated by Steven Rendall. Cambridge, MA: Belknap Press.

Population Division of the UN Department of Economic and Social Affairs. 2018. *The 2018 Revision of World Urbanization Prospects*. Available at UN DESA, www.un.org/development/desa/publications/2018-revision-of-world-urbanization -prospects.html. Accessed 27 October 2022.

Pulido, Laura. 2018. 'Historicizing the Personal and the Political: Evolving Racial Formations and the Environmental Justice Movement'. *The Routledge Handbook of Environmental Justice*, edited by Ryan Holifield, Jayajit Chakraborty, and Gordon Walker, 15–24. Abingdon: Routledge.

Sale, Kirkpatrick. 1984 [1986]. 'Mother of All: An Introduction to Bioregionalism'. *The Schumacher Lectures: Second Volume*, edited by Satish Kumar, 219–50. London: Abacus.

Santayana, George. 1905 [1910]. *The Life of Reason*. London: Archibald Constable & Co. Available at archive.org, https://archive.org/details/in.ernet.dli.2015.94934/. Accessed 21 January 2023.

Schumpeter, Joseph A. 1943 [2010]. *Capitalism, Socialism, and Democracy*. London: Routledge.

Sen, Amartya. 1999. *Development as Freedom*. Oxford: Oxford University Press.

Scheidel, Walter. 2017 [2018]. *The Great Leveler – Violence and the History of Inequality from the Stone Age to the Twenty-First Century*. Princeton, NJ: Princeton University Press.

Steffen, W., W. Broadgate, L. Deutsch, O. Gaffney, and C. Ludwig. 2015. 'The Trajectory of the Anthropocene: The Great Acceleration'. *Anthropocene Review* 2: 81–98. doi: https://doi.org/10.1177/2053019614564785

Steinbach, Susie L. 2017. *Understanding the Victorians: Politics, Culture and Society in Nineteenth-Century Britain*. 2nd edn. Abingdon: Routledge.

Stiglitz, Joseph E. 2002. *Globalization and Its Discontents*. London: Penguin.

Taylor, Jesse Oak. 2016. *The Sky of Our Manufacture: The London Fog in British Fiction from Dickens to Woolf*. London: University of Virginia Press.

Thomas, Edward. 1909. *Richard Jefferies: His Life and Work*. London: Hutchinson & Co.

Tyner, James A. 2016. *Violence in Capitalism: Devaluing Life in an Age of Responsibility*. Lincoln, NE: University of Nebraska Press.

United Nations Development Programme (UNDP). 2022. Technical Report: *Environmental Justice: Securing our Right to a Clean, Healthy and Sustainable Environment*. Available at UNDP.org, www.undp.org/publications/environmental-justice-securing-our-right-clean-healthy-and-sustainable-environment. Accessed 27 October 2022.

UN-Habitat. 2018. 'Adequate Housing and Slum Upgrading'. United Nations Human Settlement Programme (*UN-Habitat*). Available at: UN-Habitat.org, https://unhabitat.org/sites/default/files/2020/06/indicator_11.1.1_training_module_adequate_housing_and_slum_upgrading.pdf. Accessed 27 October 2022.

United Nations Human Settlements Programme. 2018. 'Population Living in Slums (% of Urban Population)'. Available at The World Bank, https://data.worldbank.org/indicator/EN.POP.SLUM.UR.ZS. Accessed 27 October 2022.

Wells, H. G. 1895 [2005]. *The Time Machine*, edited by Patrick Parrinder. London: Penguin.

White, Fred. M. 1903 [2020]. 'The Dust of Death'. *Nature's Warning: Classic Stories of Eco-Science Fiction*, edited by Mike Ashley, 41–58. London: British Library.

Wilmer, Clive. 1993 [1998]. 'Introduction'. William Morris, *News from Nowhere and Other Writings*, edited and introduced by Clive Wilmer, ix–xli. London: Penguin.

Index

The letter 'n' following a page number denotes a 'note'

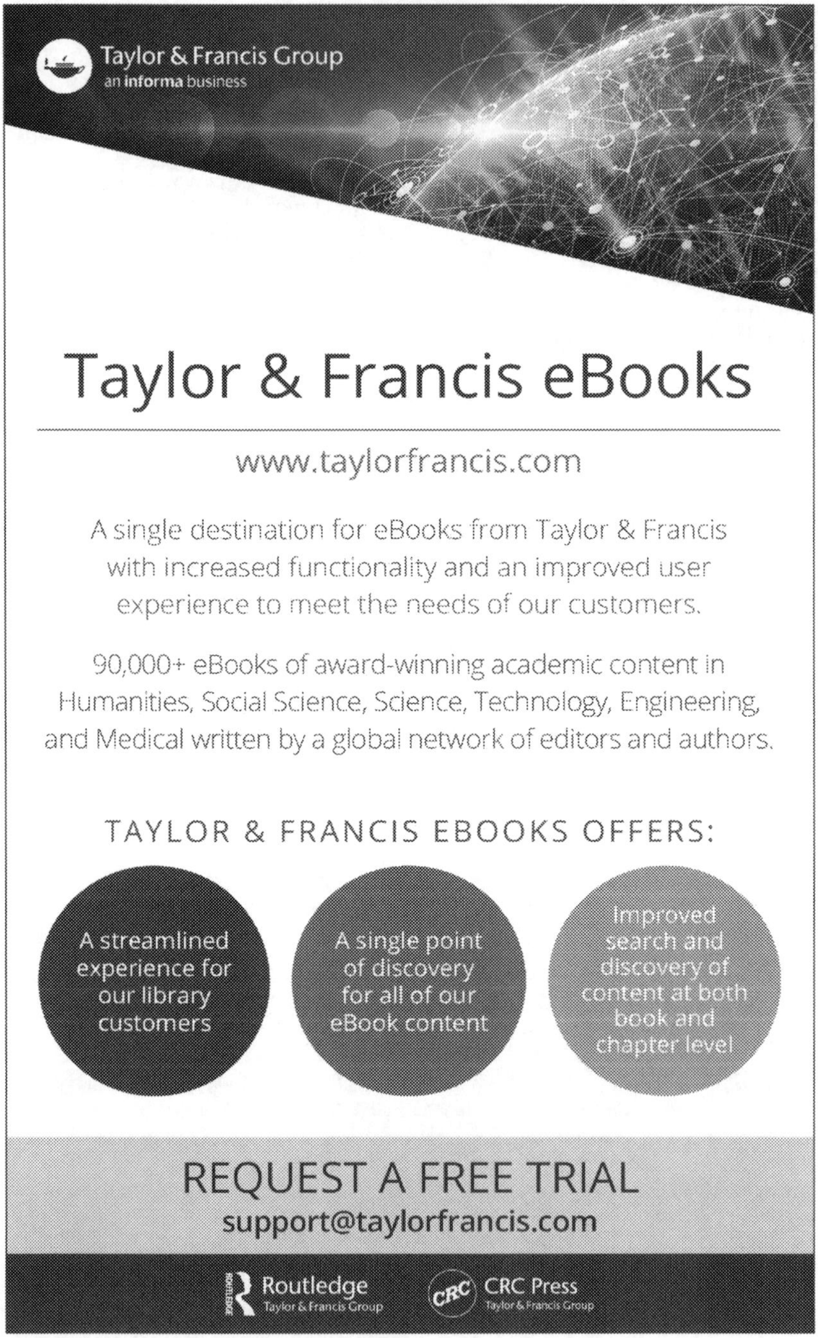

Printed in the United States
by Baker & Taylor Publisher Services